SpringerBriefs in Earth System Sciences

For further volumes:
http://www.springer.com/series/10032

Ioannis Liritzis · Ashok Kumar Singhvi
James K. Feathers · Gunther A. Wagner
Annette Kadereit · Nikolaos Zacharias
Sheng-Hua Li

Luminescence Dating in Archaeology, Anthropology, and Geoarchaeology

An Overview

Springer

Ioannis Liritzis
Laboratory of Archaeometry
Department of Mediterranean Studies
University of the Aegean
Rhodes
Greece

Ashok Kumar Singhvi
Physical Research Laboratory
Planetary and Geosciences Division
Ahmedabad
India

James K. Feathers
Luminescence Dating Laboratory
University of Washington
Washington, WA
USA

Gunther A. Wagner
Annette Kadereit
Heidelberger Lumineszenzlabor
Geographisches Institut
Universität Heidelberg
Heidelberg
Germany

Nikolaos Zacharias
Department of History
Archaeology and Cultural Resources
Management
University of Peloponnese
Kalamata, Old City Center
Greece

Sheng-Hua Li
Department of Earth Sciences
The Hong Kong University
Hong Kong
People's Republic of China

ISSN 2191-589X ISSN 2191-5903 (electronic)
ISBN 978-3-319-00169-2 ISBN 978-3-319-00170-8 (eBook)
DOI 10.1007/978-3-319-00170-8
Springer Cham Heidelberg New York Dordrecht London

Library of Congress Control Number: 2013936541

Printed on acid-free paper

Springer is part of Springer Science+Business Media (www.springer.com)

Contents

Abstract

Half a century after the publication of the first Thermoluminescence (TL) ages, the field of Luminescence Dating has reached a level of maturity. Both research and applications from all fields of archaeological science, from archaeological materials to anthropology and geoarchaeology, now routinely employ luminescence dating. The advent of optically stimulated luminescence (OSL) techniques and the potential for exploring a spectrum from mono-minerallic single grains to polymineral multi-aliquots enhanced the applicability, accuracy and the precision of luminescence dating. The present contribution reviews the physical basis, mechanisms and methodological aspects of luminescence dating; discusses advances in instrumentations and facilities, improvements in analytical procedures and statistical treatment of data along with some examples of applications across continents. The case studies review the dating of heated and solar bleached archaeological material (artefacts, sediments, rocks, rock art and buildings) that cover all periods from Middle Palaeolithic to Medieval Eras and both Old and New World archaeology. They also include interdisciplinary applications that contribute to palaeo-landscape reconstruction.

Keywords Luminescence dating · TL · OSL · Archaeology · Palaeoanthropology · Geoarchaeology · Artefacts · Monuments · Sediments · Dating · Authenticity

Chapter 1
Introduction

1.1 Introduction

The tenets of luminescence dating were worked out using thermoluminescence (TL) of archaeological pottery. The early pioneering work was led by Martin Aitken (Aitken et al. 1964) and this established TL as a robust method for the dating of pottery as well as for the authentication of heated excavation materials, pieces of art and objects related to cultural heritage. In the 1980s, the demonstration that daylight exposure could also reset luminescence signals initiated a new era of application to geological and archaeological sediments. Development of optically stimulated luminescence (OSL; Huntley et al. 1985) expanded luminescence application and today OSL forms a major fraction of luminescence research in Quaternary geology and archaeology.

Many artefacts are either too sparse or too valuable to be destructively sampled for dating or characterization. However, geological materials in archaeological contexts, such as sun bleached soil floors, overlying pebbles, tephra deposits, burnt clay, and raw obsidian, provide useful information on the dating of cultural phases, palaeoanthropology, characterization, trade exchange, and on the dating of geological events related to human evolution. Thus, geoarchaeological materials in archaeological contexts are potential chronometers (McDougall 1968).

Luminescence dating also enables age estimation of stone structures (monoliths, buildings, cairns, field walls etc.,). The material associated with the construction period is directly dated. This is in contrast to radiocarbon dating where the material dated is only associated with the archeological material. In many cases appropriate organic debris is either not available, or the association is not established unambiguously. OSL and TL date the most recent exposure of a mineral grain to daylight or heating. If in the course of construction, sediments bleached prior to deposition or a stone surface got buried or permanently shielded from further light influence, luminescence dating provides a tool to directly date the time of construction.

Luminescence is the light emitted from minerals, such as quartz and feldspar following an exposure to ionizing radiation. Luminescence arises due to the

I. Liritzis et al., *Luminescence Dating in Archaeology, Anthropology, and Geoarchaeology*, SpringerBriefs in Earth System Sciences, DOI: 10.1007/978-3-319-00170-8_1, © The Author(s) 2013

presence of defects in the mineral's crystal lattice. Ionizing radiation creates free electrons and holes in the lattice and a few of these get trapped in defects with opposite charges. The residence time of charges in the defects can range from a few seconds to Myrs. The trapped charges additionally can be excited by an external thermal or optical stimulus (i.e., heating to ~400 °C or day or laboratory light exposure of few seconds). The freed charges then wander in the crystal lattice and some of these radiatively recombine with the opposite charges at another defect site to give light. The intensity of the emitted light is proportional to the concentration of trapped charges/electrons and hence to the radiation dose. The latent luminescence signal increases till a saturation of trapped charges occurs.

Dating using luminescence is made possible by the fact that in natural archaeological and geological environment, the decay of natural radionuclides viz. potassium and radioactive decay chains viz. uranium, thorium along with cosmic radiation, provide a constant irradiation field. Therefore, the minerals in the sediment or archaeological object are irradiated at a constant rate, and hence acquire latent luminescence at a constant rate. The latent luminescence is released upon exposure to heat or light, setting the signal to zero or near zero, whence the trapping process begins anew. Events which zero the pre-existing geo- or archaeoluminescence are intentional or accidental exposure to heat (>400 °C) or exposure to daylight which provides sufficiently energetic photons to induce zeroing. Authigenic minerals that form in situ, also can be dated as these acquire their luminescence since the establishment of their crystal structure Fig. 1.1.

In the laboratory the same process is mimicked during the dating procedure. The trapped charge population can be measured by stimulating the crystal by heat (mostly up to ~400 °C) or visible (mostly blue or green light) or infrared (IR) light. These stimulations lead to release of charges some of which recombine radiatively with opposite charge carriers, thereby emitting luminescence in either or all of the ultraviolet (UV), the visible or the IR-spectrum. The intensity of this light (luminescence) is proportional to the number of recombining charges and this

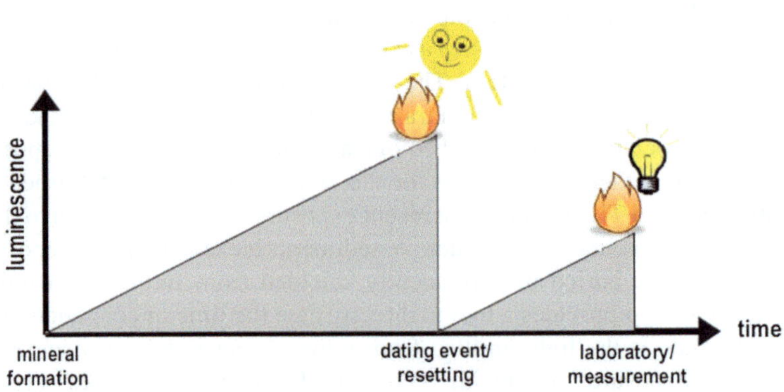

Fig. 1.1 Schematic diagram. Growth and resetting of the latent luminescence signal for dated events.

in turn is proportional to the amount of trapped charges. This fact is exploited to convert light units to dose units. Because this light (luminescence) emission is stimulated by exposure to heat/light it is called *thermally* (TSL or TL)—or *optically stimulated luminescence* (*OSL*). The intensity of the latent luminescence acquired since the last event of charge eviction is proportional to a sample's age:

$$age = total\,luminescence/annual\,rate\,of\,luminescence\,acquisition$$

Given that luminescence is proportional to the dose, the above equation can be rewritten as,

$$age = paleodose\,(D_e)/annual\,dose\,(D_T)$$

where D_e is the laboratory beta dose that induces the same luminescence intensity in the sample as emitted by the natural sample. D_T is the annual dose-rate and comprises several components of radiation that arise from the decay of natural radioactive elements i.e., alpha, beta and gamma rays along with a minor contribution from the cosmic rays (see, Chap. 3).

Artifacts that can be dated include ceramics, burned lithics, burnt bricks and soil (e.g., from hearths), and potentially unburnt stone surfaces that were exposed to light and then buried. Sediments that can be dated using luminescence include aeolian deposits (dune sands or loessic dust), paleosols, colluvial deposits, and water- and ice lain deposits (e.g. marine, fluvial, lacustrine), (Aitken 1985, 1998; Roberts 1997; Murray and Olley 2002; Liritzis 2000; Liritzis et al. 2002; Wintle 2008; Preusser et al. 2008; Fuchs and Lang 2009). For all these annealed or bleached geoachaeological materials the 'luminescence clock' is set up that starts counting the time.

Application of luminescence methods for various geological domains were reviewed in a special issue of BOREAS (2008). Application of luminescence in archaeology and anthropology were reviewed by Galbraith et al. 1999; Theocaris et al. 1997; Roberts 1997; Feathers 2003; Richter 2007. These include dating of the burial context in archaeological deposits, artifacts, such as ceramics, pot boilers, fireplaces, rock art, and wasp nests in caves (Liritzis and Galloway 1999; Liritzis et al. 1994; Roberts et al. 1997).

In the following we review the potential of luminescence dating as applied to archaeological, geoarchaeological and anthropological materials related to human and cultural evolution. Important applications, innovative instrumentation and new methodological procedures are described.

References

Aitken MJ (1985) Thermoluminescence dating. Academic Press, London

Aitken MJ (1998) An introduction to optical dating. Oxford University Press, Oxford, 280pp

Aitken MJ, Tite MS, Reid J (1964) Thermoluminescent dating of ancient ceramics. Nature 202:1032–1033

BOREAS journal (2008) 37(4). doi:10.1111/bor.2008.37.issue-4/issuetoc
Feathers JK (2003) Use of luminescence dating in archaeology. Meas Sci Technol 14:1493–1509
Fuchs M, Lang A (2009) Luminescence dating of hillslope deposits—a review. Geomorphology 109:17–26
Galbraith RF, Roberts RG, Laslett GM, Yoshida H, Olley JM (1999) Optical dating of single and multiple grains of quartz from Jinmium rock shelter, northern Australia: Part I, Experimental design and statistical models. Archaeometry 41:339–364
Huntley DJ, Godfrey-Smith DI, Thewalt MLW (1985) Optical dating of sediments. Nature 313:105–107
Liritzis I (2000) Advances in thermo- and opto-luminescence dating of environmental materials (sedimentary deposits): Part I: Techniques Global Nest: The Int J 2(1):3–27, and Part II: Applications, Global Nest: 2(1):29–49
Liritzis I, Galloway RB, Theocaris PS (1994) Thermoluminescence dating of ceramics revisited: optical stimulated luminescence of quartz single aliquot with green light emitting diodes. J Radioanal Nucl Chem Lett 188:189–198
Liritzis I, Galloway RB (1999) Dating implications from solar bleaching of thermoluminescence of ancient marble. J Radioanal Nucl Ch 241(2):361–368
Liritzis I, Galloway RB, Hong D, Kyparissi-Apostolika N (2002) OSL dating of three prehistoric ceramics from Theopetra cave, Greece: a case study. Mediterr Archaeol Archaeometry 2(2):35–43
McDougall DJ (ed) (1968) Thermoluminescence of geological materials academic press
Murray AS, Olley JM (2002) Precision and accuracy in the optically stimulated luminescence dating of sedimentary quartz. Geochronometria 21:1–16
Preusser F, Degering D, Fuchs M, Hilgers A, Kadereit A, Klasen N, Krbetschek M, Richter D, Spencer J (2008) Luminescence dating: basics, methods and applications. Eiszeitalter & Gegenwart, Hannover, Quaternary Sci J 57(1–2):95–149
Richter D (2007) Advantages and limitations of thermoluminescence dating of heated flint from Paleolithic sites. Geoarchaeology 22:671–683
Roberts RG (1997) Luminescence dating in Archaeology: from origins to optical. Rad Meas 27:819–892
Theocaris P, Liritzis I, Lagios E, Sampson A (1997) Geophysical prospection and archaeological test excavation and dating in two Hellenic pyramids. Surv Geophys 17:593–618
Wintle AG (2008) Fifty years of luminescence dating. Archaeometry 50(2):276–312

Chapter 2
Luminescence Dating Protocols and Dating Range

Luminescence dating comprises estimation of paleodose, or rather the equivalent dose (D_e), and annual dose rate (D_T). These are discussed below.

The estimation of D_e can be done using multiple aliquot (MA), single aliquot (SA) or single grain (SG) techniques and in each case additive dose or regenerative dose procedures are used. In the additive dose procedures, several laboratory doses of varying magnitude are given additionally on top of the natural dose of a sample, on several identical subsamples of a natural sample (so called aliquots). The luminescence signal from the natural dose, as well as the natural plus added doses is plotted against the added doses (zero added dose for the natural) and the relation is fitted with a linear or exponential curve, which describes the growth of the luminescence signal with increasing dose (growth curve). The additive growth curve is extrapolated to the dose axis to provide an estimate of the equivalent dose. In the regeneration method, the natural signal is bleached first and then doses are added to construct a luminescence vs. dose growth curve. The natural signal is then interpolated on to this regenerated growth curve to estimate the equivalent dose.

The techniques are further discussed below.

2.1 Multiple Aliquots Additive Dose

In this method the additive technique is applied to multiple aliquots. This means that each dose point of an additive growth curve is represented by the (mean) luminescence from several aliquots. For reliable results, it is important to ensure that all aliquots are identical and that the extrapolation is both realistic in terms of the underlying physical mechanisms and accurate. Jain et al. (2003) review all the normalization procedures to normalize aliquots and evaluate their efficacy to produce identical subsamples. Felix and Singhvi (1997) review the extrapolation procedures and optimization of the measurement protocols in terms of errors. The advantage of multiple aliquot additive dose (MAAD) is that it averages the luminescence signal over several thousand grains and hence provides a mean age

I. Liritzis et al., *Luminescence Dating in Archaeology, Anthropology, and Geoarchaeology*, SpringerBriefs in Earth System Sciences,
DOI: 10.1007/978-3-319-00170-8_2,

for an ensemble of grains. MAAD is suitable for samples for which heterogeneity in zeroing may be excluded (e.g. heating or daylight bleaching at grain level) (Fig. 2.1a, b). For heterogeneously zeroed samples, however, non judicious use of MAAD methods can lead to erroneous results.

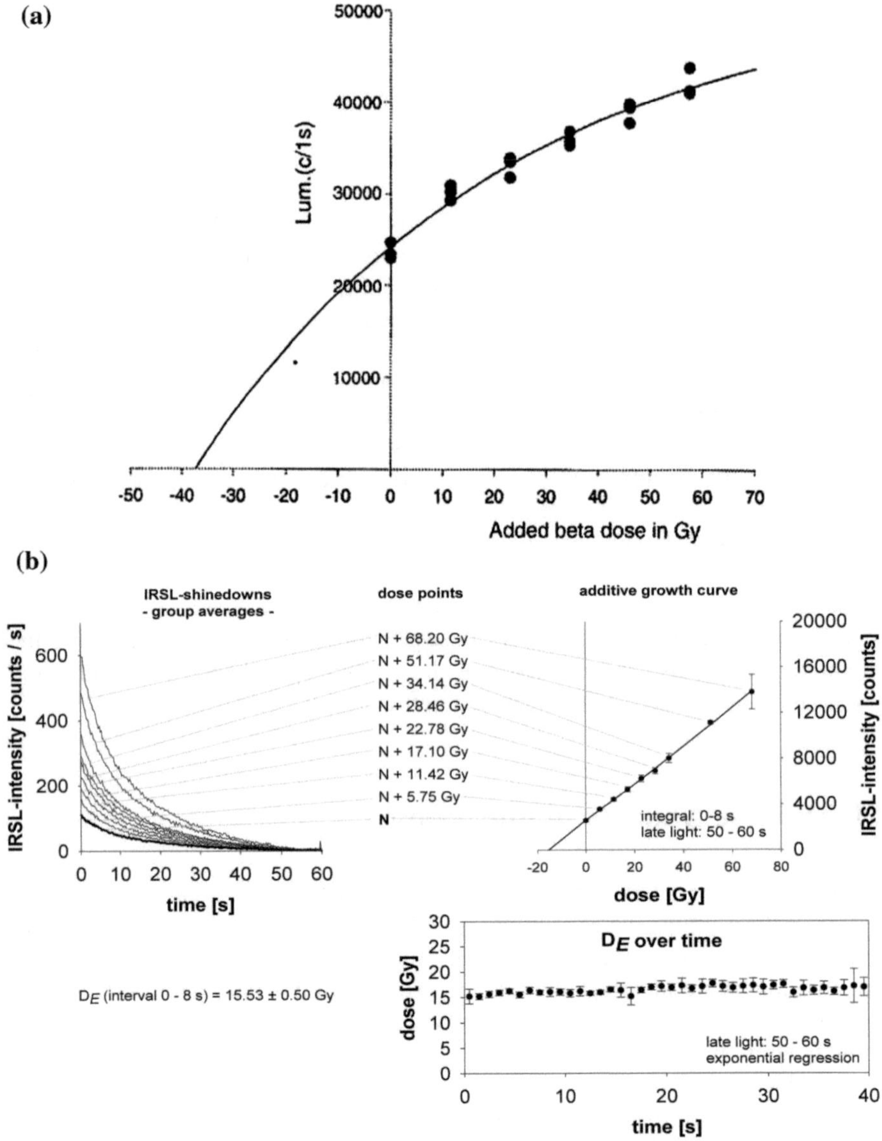

Fig. 2.1 Illustration of MAAD-protocol for (**a**) growth curve of ceramic quartz (3 aliquots per dose) fitted with a single exponential function (Liritzis et al. 2002), (**b**) IRSL fine-grains with shine-down curves (*upper left*), additive growth-curve (*upper right*) and DE-plateau test (*lower right*) (Kadereit et al. 2007)

2.2 Single Aliquot and Single Grain Techniques

In introducing OSL dating, Huntley et al. (1985) suggested that it should be possible to make sufficient measurements on a single aliquot to allow a D_e determination. Duller (1991) developed a single aliquot method for D_e-determination by administering additive doses to potassium feldspar extracts. This single aliquot additive dose (SAAD) technique requires correction for sensitivity change during read outs, (Liritzis et al. 1997, 2001, 2002). In SAAD a single aliquot (disc) is measured with consecutively administrering beta doses and reading the OSL by short shining from diodes at certain wavelength. The signal growth is fitted by appropriate functions. A single aliquot regeneration (SAR) method developed for quartz is now widely used for the dating of sediments (Murray and Wintle 2000). The SAR protocol—given dose Di, preheat, OSL reading (Li), test dose Dt, preheat, measured OSL (Ti), repeated steps—takes into account possible sensitivity changes during the construction of the regeneration growth curve. Sensitivity may change from repeated irradiation, preheating and OSL stimulation of an aliquot. The development of automated luminescence readers with capabilities of carrying out complete measurement sequences including stimulation and read out of the luminescence signal as well as irradiation and preheat of aliquots without the need to physically move the samples, has made it possible to undertake all the measurements necessary for D_e determination in a sequential way. Given that a large number of independent measurements can be made on many single aliquots in principle it is possible to improve the precision of the paleodose measurement to any desired level, by simply increasing the number of measurements (Fig. 2.2a, b).

When using the SAR protocol, several tests and checks are required to ensure reliable D_e values (Wintle and Murray 2006; Singhvi et al. 2010, 2011). These checks include: (a) making sure the sensitivity correction is consistent for identical doses (recycling test), (b) testing for any build up of dose from preheating (recuperation test), (c) testing quartz separates for feldspar contamination, (d) optimizing the preheat (by plateau tests), (e) dose recovery of a known dose, (f) plotting D_e against the stimulating time to test for partial bleaching, and (g) crystal sensitivity change. If the sample fails any of the test/checks, the data are discarded. A logical extension of single aliquot methods is the use of single grains. This permits age determination on individual grains and hence provides the best resolution for understanding the bleaching history of the samples. On the flip side is the fact that radiation dose at single-grain scale can be heterogeneous, requiring precautionary measures (Mayya et al. 2006; Chauhan et al. 2011). In the SAR protocol for quartz grains, it is an underlying assumption that the measured OSL signal is dominated by a signal that is most easily bleached, called the fast component. Isolation of the fast component can be done either mathematically by curve fitting (Murari 2007) or experimentally by using linearly modulated OSL (LM-OSL) where the stimulation power is increased linearly and the OSL is measured (Singarayer and Bailey 2004). The LM-OSL components can also be separated by curve fitting (Kitis et al. 2002; Li and Li 2006a). We refer to

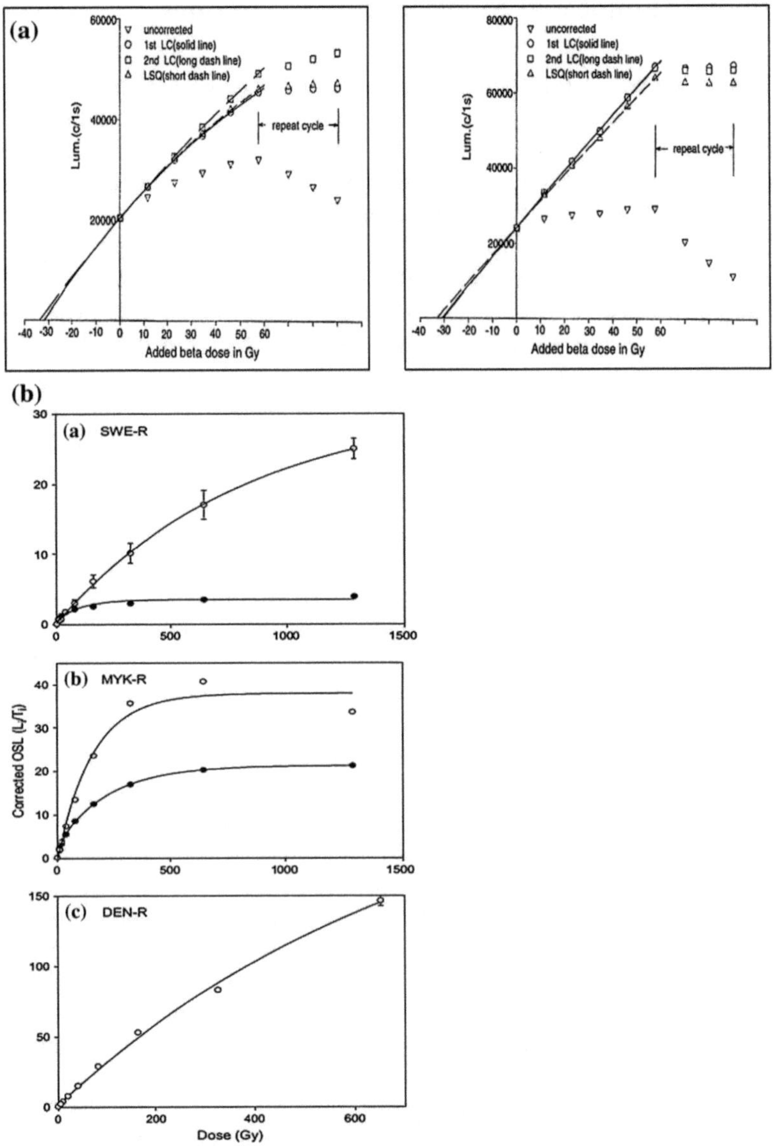

Fig. 2.2 Plots illustrate **a** the single aliquot additive dose (*SAAD*) on ceramic following Duller (1991) shining and preheating crystal sensitivity correction labelled "1st LC", the Duller (1994) correction labelled "2nd LC", and the iterative least squares method of Galloway (1996) and Liritzis et al. (1997) labelled "LSQ". The points from each correction procedure are fitted by a saturating exponential curve, which is extrapolated to the equivalent dose value. Preheating was for *a* 1 min at 220 °C and for *b* 5 min at 220 °C (Liritzis et al. 2002), **b** single aliquot regenerated (*SAR*) technique, curves for single slices of rock samples *a* Swedish ultramafic, *b* Mykonos granite and *c* Danish quartz metamorphic. The *filled circles* represent the sensitivity corrected blue OSL response and the *open circles* the IR response. The growth curves have been fitted with an expression of the form $y = \alpha\,(1 - e^{-bx})$, Li/Ti is defined the ratio of IR readings during OSL SAR protocol, where Li and Ti are natural and regenerated points as a function of laboratory dose Di (Vafiadou et al. 2007)

interesting papers by Choi et al. (2003) and Li and Li (2006b) for other methodological possibilities using LM-OSL, such as the detection of an ultrafast component and influence of medium bleaching components. It is important to notice that isolation of components is desirable as different components may have different dose response curves.

An important aspect is the calibration of beta dose rate in the laboratory. The net dose absorbed depends on the type of radiation and stopping power of the irradiated material and the backscattering from the mounting disc on which the grains are kept. Thus, the dose rate delivered by the same source to fine grain quartz mounted on aluminium disc is different from fine grain feldspar on steel discs. Similarly the dose rate to quartz and feldspar grains of identical size are different by ~7 % on account of differences in the stopping power. Also different sizes of quartz grains also absorb different amounts of doses. Thus, for example, the dose to fine grained (4–11 m) quartz mounted on aluminium disc could be lower by ~25–30 % compared to 100 μm quartz grains mounted on stainless steel (SS) discs. Therefore, beta dose calibration needs to be established for each mineral type and grain size and measurement condition (e.g. material of the disc on which the grains are mounted). The Risoe National Laboratory/DTU in Denmark now provides standard quartz for calibration purposes with nominal charges (Kadereit and Kreutzer 2013).

Table 2.1 gives the protocols used for D_e determination. The single aliquot regeneration (SAR) protocol (2000) is now the most widely applied technique worldwide and the recent revision of this protocol (Singhvi et al. 2011) provides a robust way to estimate ages. A variety of further modifications exist, e.g. pulsed optical stimulation which allows the quasi-simultaneous detection of several OSL/IRSL-emissions; thermally transferred OSL for very old and pre-dose OSL for very young samples, respectively (for further details see text). As 'old fashioned' multiple aliquot protocols often deliver reliable results they are still widely applied when material availability is not a problem and however small aliquots to check heterogeneous bleaching cannot be prepared routinely (e.g. from fine-grained loess). Also, TL-dating might be more appropriate for heated objects than OSL (e.g. flint) though a concordence between TL and OSL results has been established over and again. Furthermore, novel TL-techniques have been developed even for sediments with the aim to extend the upper dating limit for older samples for which an unbleachable luminescence residuum does not lead to significant age-overestimation (Vandenberghe et al. 2009).

2.3 New Developments

Luminescence dating based on measurement of the fast component of the OSL quartz signal plays a major role for dating late Quaternary sediments though limited to ca 100–200 ka because of saturation. A thermally-transferred OSL (TT-OSL) signal is observed when quartz is heated to ~200–300 °C after an

Table 2.1 Schematic overview of the basic stimulation, detection and protocol types of luminescence dating

Aliquot type	Protocol	Stimulation/detection				
		TL/UV to red[16]	OSL-blue/UV	OSL-green/ blue to UV	IRSL/UV[7], blue, yellow	IR-RF/IR [865 nm][19]
Multiple aliquots	Additive dose	Ceramic (q, f, pfg) R-B[1]: sediments (q, f, pfg) Calc. stone surface (c) [520 nm][15]		Sediments (q, f)[4]	Ceramics (f) Sediments (f)	
	Regeneration (slide)[21]	Ceramic (pfg)[20] Heated flint (q) Sediments (q)				
Semi-single aliquots	SARA[2]	Burned stone (q, f)[2a] Ceramic and brick (q)[2a]		Ceramics (q)[2a, 2b] Brick (q)[2a] Burned stone (q)[2b] Sediment (f)[2b]		
	SA additive[3]					
Single-aliquot	SAR[11]	Heated flint (q) [620 nm][10] Slag (q) [620 nm][9]	Ceramics (q) Sediments (q) Calc. stone surface (q)[14]			
	IRSL (and post-IR/ IRSL)-SAR				Sediment (f)	Sediments > 20 ka (f)

(continued)

Table 2.1 (continued)

Aliquot type	Protocol	Stimulation/detection				
		TL/UV to red[16]	OSL-blue/UV	OSL-green/ blue to UV	IRSL/UV[7], blue, yellow	IR-RF/IR [865 nm][19]
Single-grain	SG-SAR (includes OSL, post IR-OSL, and IRSL)			Sediments (q)[12] Mortar (q)[13]	Sediment (f)	
ROI[8]	HR-OSL SAR			Granitic and limestone (with quartz) stone surface (f)[5]	Granitic and limestone (with quartz) stone surface (f)[5]	

Generally, the development was from TL (1950s) to OSL (1985) and to IRSL (1988) and to RF (1999); from the dating of ceramics to sun bleached sediments (1978); from multiple-aliquot to single-aliquot (1990s) and to single-grain dating (1999) and to the dating of small parts (ROIs) of intact stone surfaces (2002). These include limestone and sandstone rocks from masonry that include quartz
f = feldspar
q = quartz
pfg = polymineral fine-grain
1 A special technique to subtract the unbleachable background of a TL-signal (Wintle and Huntley 1980; Berger et al. 1987), necessary for sediment dating
2 A regenerative protocol with few additively dosed aliquots introduced by Mejdahl and Botter-Jensen (1994[2a], 1997[2b]), but hardly or not in use any longer
3 Developed by Duller (1991[3a], 1994[3b]) for coarse-grain potassium feldspars from sediments, using additional monitor aliquots to record changes of an aliquot during a dating procedure; on loose grains extracted from masonries (see Liritzis et al. 2010b)
4 Huntley et al. (1985) using an argon laser on aeolian quartz
5 Greilich et al. (2002, 2005) analyzing feldspar from granitic stone surfaces from Peruvian geoglyphes and a castle in S-Germany (slice technique for sample preparation)
6 Banerjee et al. (2001)
7 The UV-emission is often regarded as especially unstable, showing unwanted a thermal signal loss (fading) (Lang 1996)
8 Range of interest: an area within a stone surface with a bright luminescence signal and good dating properties, usually corresponding to a 'single grain', yet still preserved in its original position of the mineral assemblage and dose field
9 Haustein and Krbetschek (2002)
10 Richter and Krbetschek (2006)
11 Murray and Wintle (2000) introducing the basic version of the SAR protocol most luminescence laboratories in the world refer to nowadays
12 Olley et al. (1999)
13 Jain et al. (2004)
14 Liritzis et al. (2010) dating quartz from surfaces of calcareous schist blocs (powder technique for sample preparation)
15 Liritzis (1994), Liritzis et al. (1997) dating calcite from the surface of marble blocs by the use of a partially bleached TL MAAD technique (powder technique)
16 Detected range of emission given in square brackets
17 For heterogeneously bleached samples, so called 'small aliquots' (Olley et al. 1998) are used for coarse grains, with only few hundreds to few tens of grains, instead of thousands of grains per aliquot
18 Trautmann et al. (1999a, b) introducing the new technique under the name radioluminescence (RL)
19 Emission from luminescence traps of feldspar due to radiative electron capture, not from luminescence centres
20 Feathers (2009)
21 Prescott et al. (1993)

optical bleach. As the comparatively small OSL signal saturates at far higher radiation doses it offers the potential to date sediments back to, and even more, than, 1 million years (1 Ma). The optical and thermal behaviour of the TT-OSL signal has been characterised by Wang et al. (2006a, b) who showed that it consists of two signals: a recuperated OSL (ReOSL) signal and a basic-transferred OSL (BT-OSL) signal. A SAR protocol had been developed (Wang et al. 2007) with loess deposits dated successfully back to the Brunhes-Matuyama time-marker horizon at 780 ka (see, also, Porat et al. 2009). Though under continuous testing, the method has up to now provided sensible luminescence ages from several case studies and a good basis for understanding the applicability and limitations of the method is now available. Athanassas and Zacharias (2010) carried out a preliminary study on raised beaches in southern Greece; their bleaching experiments showed that exposure to daylight in Athens for half an hour reduced their TT-OSL signal to less than 15 % of its initial signal but the low TT-OSL signal obtained was a major impediment and currently restricts application to samples with very bright luminescence emission. Jacobs et al. (2011) produced ages for raised marine sediments from Quaternary landforms along the Cape coast of South Africa that are consistent with an interglacial high sea level (MIS 11) around 400 ka. One major research outcome from that study is that both TT-OSL signals are much less sensitive to light than previously thought, thus only well-bleached samples should be considered for TT-OSL dating, which restricts the range of possible applications of this method. Furthermore, the reliability of TT-OSL and the age range over which it can be applied still needs more work via the analysis of samples with good independent age control.

The SAR protocol has been used for both quartz and feldspar minerals. But often quartz samples possess low sensitivity and low saturation dose that limits the dating range to about 200 ka. On the other hand, feldspar despite their ten fold higher luminescence sensitivity and higher saturation dose, may suffer from athermal loss of luminescence signal (called anomalous fading) and hence may provide underestimated ages. This undesirable feature may be minimized by selecting more stable luminescence emissions and appropriate storage and preheat procedures, as e.g. shown for loess samples from southern Germany by Lang et al. (2003). Several other methods have now been developed to enable correction/circumvention of the fading component in feldspar luminescence (Huntley and Lamothe 2001). Three recent developments merit a mention.

The first is the use of the IRSL signal from K-feldspars using an isochron for different grain sizes as proposed by Li et al. (2007, 2008a, b). This requires the sample to contain a wide range of grain sizes typically in the range from 90 to 250 μm from which K feldspar is extracted using density separation with heavy liquids. Under the assumption that the grains were well-bleached, such grains receive variable amounts of internal doses based on their sizes (Li et al. 2008a) bearing in mind that this is independent of wether they were sufficiently bleached or not. The D_e values increase with grain size, due to an increased component of internal dose. A plot of D_e values as a function of the internal dose rate (contributed from the potassium and rubidium content of the K-feldspar grains) enables creation of an isochron plot. An effective external dose contribution is then calculated for each grain size after consideration of the attenuation of the radiation by the grain size. The *isochron IRSL* (*iIRSL*) age is then obtained as the difference between the slopes of plots of D_e and external dose attenuation against the internal dose rate, calculated for the grain sizes. The iIRSL method may overcome age underestimation due to anomalous fading. Also, since the isochron method (iIRSL) is reliant on only the internal dose rate, it overcomes problems related to: (1) changes in past dose rate due to post-depositional migration of radionuclides, (2) changes in water content as waterlain sediments dry out, (3) spatial heterogeneity in the gamma dose rate, (4) uncertainties in the cosmic ray dose rate during the period of sample burial.

A second more recent effort uses IRSL measurements at elevated temperatures, e.g. 290 °C, after the IR signal is read out at 50 °C (Ankjaergaard et al. 2010; Buylaert et al. 2009; Jain et al. 2011). The post-IR luminescence signal measured at elevated temperature has a smaller fading rate than the signal measured at 50 °C. Recently, it was proposed that a non fading signal can be achieved by using several elevated temperatures and looking for a plateau of De values in terms of stimulation temperature (Li and Li 2011). The plateau region is considered stable. Though physically sound, the efficacy of this proposal needs testing and several studies are in progress and some have shown good results (Biswas et al. 2013).

In a third approach, presence of a signal with a stability of only 50 ka is considered and based on this it is surmised that the signal is lost due to loss of charge luminescent centres. A sensitivity based correction for a sample as received (with decayed centres) and the sensitivity of the same sample rejuvenated via a large dose (to mimic geological dose) and day light exposure is compared. The ratio provides a correction factor for the paleodose. The use of such sensitivity ratio has yielded sensible results (Biswas et al. pers. comm.).

Another novel technique, though sparingly used by researchers, is the *infrared radiofluorescence* (*IR-RF*) [(Trautmann et al. 1999a, b, 2000; Schilles and Habermann 2000); earlier terminology: radioluminescence (RL)]. Although related to other trapped-charge techniques, it does not use the signal resulting from recombination of charge after trap eviction. Rather it monitors an IR-signal around 865 nm that is emitted by electron capture at luminescence traps during ionizing laboratory irradiation. This way, radiofluorescence is not influenced by the possible problems related to lifetimes (fading), electron competition and other features concerning luminescence recombination centres, but refers only to the luminescence traps. For dating applications, a single-aliquot regenerative procedure (IR-SAR) is used (Krbetschek and Erfurt 2003a). As for the classical SAR-protocol, first the natural signal is measured, with RF intensity denoting the 'filling level' of the luminescence traps; thereafter the RF-signal accompanying the filling of the traps up to saturation is monitored during laboratory irradiation. Finally, the sample is bleached and an RF-curve of a regenerating sample is recorded. Correction of the additive and the regenerative curve is at times necessary due to possible sensitivity changes. So far, studies of IR-RF have concentrated on feldspar from sediments (e.g. Erfurt and Krbetschek 2003b; Krbetschek et al. 2008).

The RF-signal bleaches faster than a TL-signal but less rapidly than an OSL-signal. The dating range is from ca. 20 ka to several hundred ka, depending on the natural dose rate. Apart from the fact that SA-protocols on coarse-grains allow thorough scrutinizing of insufficient bleaching, an unbleachable residual RF signal is less problematic for older samples. The method therefore enables a significant extension of the upper age limit in luminescence dating. In one example, IR-RF dating was applied to coarse-grain potassium feldspar extracts from sediment archives containing Palaeolithic Neanderthal hunting sites on lake-shore environments at Neumark-Nord 2 in E-Germany. These were dated to an interstadial at around 90 ka (layer Neumark-Nord 2.0) and to the last interglacial at around 120–130 ka (layer Neumark-Nord 2.2), respectively (Strahl et al. 2010).

Recently, IR-RF-dating of potassium feldspar grains from sediment layers at the type-site of Homo heidelbergensis at Mauer in southern Germany (where a lower jaw with teeth of the oldest central European considered as the ancestor of Neanderthals in Europe was found) helped to constrain the age of the hominin to the marine isotope stage (MIS) 15 (Wagner et al. 2010). Mean IR-RF ages from small aliquots from two samples of fluvial sand from immediately below and above the gravelly find layer were 607 ± 55 and 603 ± 56 ka, respectively. Successful numeric dating of Pleistocene hominins is important in view of the numerous theories about their migration history out of Africa and from southern Europe northward, and the possible replacement of earlier population groups by later ones.

2.4 Instrumentation

Conceptually, the measurement of TL and OSL is simple. It needs: (1) a stimulation source that provides heat or light to the sample at a known intensity and rate and does so with high reproducibility (to better than 0.1 %); (2) a light detection source, normally a photomultiplier (PM) tube with appropriate filters to discriminate against the stimulation light (in the case of OSL and IRSL) and black body radiation (in case of TL) and to optimize the colour of the signal and thus extract a specific luminescence emission for dating. These parameters have been standardized by the two automated systems now available, from Risø National Laboratory, Denmark and Daybreak Nuclear and Medical Systems, USA, The most recent development, is Freiberg Instruments, Germany, however its instruments are not yet to be used routinely. These systems are also equipped with a beta irradiation source for onplate irradiation and thereby enable measurements of several aliquots in a programmed manner. These systems have custom made software for a speedy reduction of data for data analysis and age calculation.

Most commonly, the excitation light intensity is kept constant i.e. to within a fraction of percent or less, and the stimulated luminescence is measured under various constant stimulation termed as continuous wave (CW OSL). Under this mode, read out can be only at wavelengths other than the stimulating wavelength, i.e. usually at shorter wavelength to avoid unwanted noise. It produces a decay curve (cf. Fig. 2.1b, upper left), which is usually a composite of several decay curves, each attributable to different traps.

Mathematical deconvolution enables isolation of these curves so that the best suited component can be used (Murari 2007). Typically quartz OSL decay is described by 3 and sometimes up to 7 components. Feldspar IRSL decay may also comprise one or several components but this is still a matter of discussion. For CW OSL rigorous optical filtering is needed to ensure that stimulation light does not enter the detection system, given that the stimulation intensity is typically 10^{16} times stronger compared to the stimulated luminescence from the sample. Some typical filter combinations are given in Table 2.2. The use of filters in the detection channel also results in loss of light due to reflection at each of the filter surfaces and the larger distance between the sample and the photomultiplier. Separating stimulation and emission light is not necessary with pulsed OSL (POSL) for which short pulses of sub millisecond duration are given and the light is measured in between the pulses. The photomultiplier is gated and hence during the stimulation it remains inactive and is not damaged by stimulation light. This approach is very useful for samples with low photon yield. An advantage of pulsed technique is that a short illumination can be used for OSL read out without depleting the signal much (<1 %). A further advantage is the ability to measure luminescence lifetimes, the time between excitation and emission. It has been found that quartz lifetimes are much longer than feldspar lifetimes, so that selection of a proper pulse width can isolate quartz from feldspar (Ankjaergaard et al. 2010). Longer lifetime components in feldspars also appear to suffer less from anomalous fading (Jain and Ankjaergaard 2011). A third measurement mode is linear modulation OSL (LM OSL) for which the intensity of light is increased in a linear manner. This permits a better segregation of the

Table 2.2 Typical conditions used for measurements by Boetter-Jensen (2000), Yukihara and Mckeever (2011)

Stimulations	Filters	Transmission window
TL-heating	UV Schott BG9 + Hoya 340/BG 39 + Schott UG 11	270–380
TL-heating	Blue Schott BG39 + Corning 7–58 or 7–59	340–480 nm
TL-heating	Red Komar IU	
OSL-Blue Light 470 nm LED	UV Schott BG39 + Hoya 340 nm	280–340 nm
OSL-Green Light		
532 nm laser of Green diode		
IRSL LED	Blue Schott BG39 + Corning 7–59	340–480 nm
880 nm		
940 nm		
	Red Komar IU (will fill these later)	650–750 nm

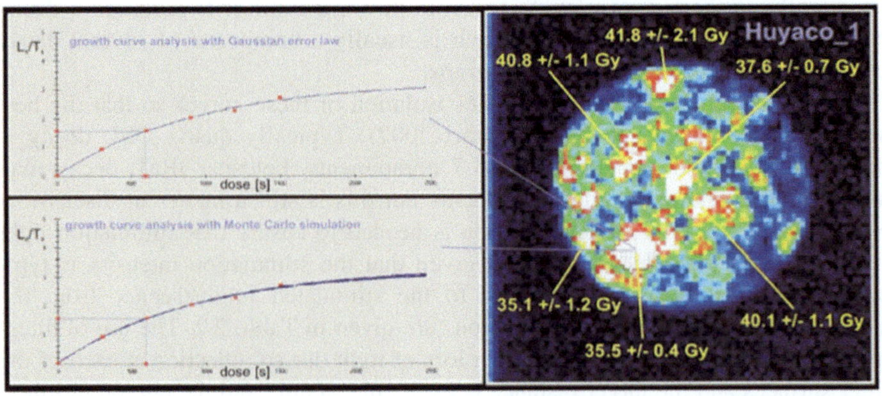

Fig. 2.3 Illustration of SAR-protocol for HR-OSL-dating on feldspar minerals within a granitic surface of a cobble. De determination for one aliquot (82 × 82 areas, each area 100 μm side length) of the sampled boulder. Luminescence intensities are presented in *false colours*. Low intensities are represented by *dark colours* (*black* and *dark blue* denoting background noise), and the brightest luminescence signal have a *white filling* (Kadereit et al. 2009)

components, allowing, in particular, the isolation of faster components for dating applications.

In TL the heating is generally done using resistive heating and is achieved by passing a large current through a kanthal or nichrome heater strip with feedback control to ensure faithful reproduction of the linear heating rate. Normally the heating is done with flowing inert gas, nitrogen or argon, to eliminate spurious luminescence due to reactions with oxygen or water vapour, and also to serve as an efficient mechanism for heat conduction from the heater plate to the sample, ensuring that the sample faithfully follows the temperature of the heater plate for the entire duration (Table 2.2).

Another innovation has been the use of CCD chips and variants for imaging the sample surface and providing spatially resolved luminescence. These detectors provide high detection efficiency close to that of a PM tube. The basic philosophy is to stimulate a rock slice (~2 mm or less thick, ~10 mm in diameter) and measure luminescence of rocks as received as well as after laboratory irradiations and to then use from the images selected pixels for the analysis. Key issues in these applications are optical depth and material density of the sample. Optical depth determines the depth in the sample from which the luminescence reaches the detection system and the density determines the depth to which electrons penetrate the sample. The new spatially resolved dating technique (high resolution OSL resp. HR-OSL) was successfully applied to the dating of stone surfaces from a stone wall of the medieval castle of Lindenfels (Odenwald, 12th century AD), Germany, and from the pre-Columbian Nasca lines (geoglyphs) around Palpa, in southern Peru (200 BC–600 AD) (Greilich et al. 2002, 2005, 2007; Greilich and Wagner 2009) as well as, a stone from a fluvial deposit from that area (Fig. 2.3).

References

Athanassas C, Zacharias N (2010) Equivalent dose estimation in coarse quartz from Pleistocene coastal sediments of south Greece using single-aliquot TT-OSL. Quat Geochronology 5(1):65–75

Ankjaergaard C, Jain M, Thomsen KJ, Murray AS (2010) Optimising the separation of quartz and feldspar optically stimulated luminescence using pulsed excitation. Radiat Meas 45:778–785

Banerjee D, Murray A.S, Bøtter-Jensen, L, Lang A (2001) Equivalent dose estimation using a single aliquot of polymineral fine grains. Radiation Measurements 33:73–94

Berger GW, Clague JJ, Huntley DJ (1987) Thermoluminescence dating applied to glaciolacustrine sediments from central British Columbia. Canadian J Earth Sci 24:425–434

Biswas RH, Williams MAJ, Raj R, Juyal N, Singhvi AK (2013) Methodological studies on luminescence dating of volcanic ashes. Quaternary Geochronology (in press) http://dx.doi.org/10.1016/j.quageo.2013.03.004

Bøtter-Jensen L (2000) Development of optically stimulated luminescence techniques using natural minerals and ceramics, and their application to retrospective dosimetry. Riso-R-1211 (EN) 74–85

Buylaert JP, Murray AS, Thomsen KJ, Jain M (2009) Testing the potential of an elevated temperature IRSL signal from K-feldspar. Radiation Measurements 44:560–565

Chauhan N, Singhvi AK (2011) Distribution of SAR paleodoses due to spatial heterogeneity of natural beta dose. Geochronometria 38:190–198

Choi JH, Murray AS, Cheong CS, Hong DG, Chang HW (2003) The resolution of stratigraphic inconsistency in the luminescence ages of marine terrace sediments from Korea. Quat Sci Rev 22:1201–1206

Duller GAT (1991) Equivalent dose determination using single aliquots. Nucl Tracks Radiat Meas 18:371–378

Duller GAT (1994) Luminescence dating of sediments using single aliquots: new procedures. Quat Geochronol 13:149–156

Erfurt G, Krbetschek MR (2003a) IRSAR: a single-aliquot regenerative-dose dating protocol applied to the infrared radio fluorescence (IR-RF) of coarse-grain K feldspar. Ancient TL 21:21–28

Erfurt G, Krbetschek MR (2003b) Studies on the physics of the infrared radioluminescence of potassic feldspar and on the methodology of its application to sediment dating. Radiation Measurements 37:505–510

Felix C, Singhvi AK (1997) Study of non-linear luminescence-dose growth curves for the estimation of paleodose in luminescence dating: results of Monte Carlo simulations. Radiat Meas 27:599

Feathers JK (2009) Problems of ceramic chronology in the Southeast: Does shell-tempered pottery appear earlier than we think? American Antiquity 74:113–142

Galloway RB (1996) Equivalent dose determination using only one sample: alternative analysis of data obtained from infrared stimulation of feldspars. Radiation Measurements 26:103–106

Greilich S, Wagner GA (2009) Light thrown on history: the dating of stone surfaces at the geoglyphs of Palpa using OSL. In: Reideland M, Wagner GA (eds) New technologies for archaeology, natural sciences in archaeology. Springer, Berlin, pp 271–283 Chapter 16

Greilich S, Glasmacher UA, Wagner GA (2002) Spatially resolved detection of luminescence: a unique tool for archaeochronometry. Naturwissenschaften 89:371–375

Greilich S, Glasmacher GA, Wagner GA (2005) Optical dating of granitic stone surfaces. Archaeometry 47(3):645–665

Haustein M, Krbetschek MR (2002) The red thermoluminescence of quartz and its application in dating archaeometallurgical slags. Radiation Protection Dosimetry 100:375–378

Huntley DJ, Lamothe M (2001) Ubiquity of anomalous fading in K-feldspars and the measurements and correction for it in optical dating. Candian J Earth Sci 38:1093–1106

Huntley DJ, Godfrey-Smith DI, Thewalt MLW (1985) Optical dating of sediments. Nature 313:105–107

Jacobs Z, Roberts RG, Lachlan TJ, Karkanas P, Marean CW, Roberts DL (2011) Development of the SAR TT-OSL procedure for dating Middle Pleistocence dune and shallow marine deposits along the southern Cape coast of South Africa. Quat Geochronology 6:491–513

Jain M, Boetter-Jensen L, Singhvi AK (2003) Dose evaluation using multiple aliquot quartz OSL: test of methods and a new protocol for improved accuracy and precision. Radiat Meas 37:67–80

Jain M, Murray AS, Bøtter-Jensen L (2004) Optically stimulated luminescence dating: How significant is incomplete light exposure in fluvial environments Quaternaire 15:143–157

Jain M, Ankjærgaard C (2011) Towards a non-fading signal in feldspar: insight into charge transport and tunnelling from time-resolved optically stimulated luminescence. Radiation Measurements 46:292–309

Kadereit A, Wagner GA, Corvinus G (2007) OSL/IRSL fine-grain dating: a preliminary chronology of quaternary slope wash and alluvial deposits from the Dun valleys of Dang and Deokhuri in western Nepal. In: Freund G, Reisch L (eds) Prehistoric cultures in Nepal from the early palaeolithic to the neolithic and the quaternary geology of the Dang-Deokhuri Dun Valleys, vol I. Harrassowitz Verlag, Wiesbaden, pp 372–379 Appendix II

Kadereit A, Kreutzer S (2013) (in press) Risø calibration quartz—A challenge for β-source calibration. An applied study with relevance for luminescence dating. Measurement. doi: 10.1016/j.measurement.2013.03.005

Kadereit A, Greilich S, Woda S, Wagner GA (2009) Cold light from the sediments of a hot desert: how luminescence dating sheds light on the landscape development of the north-eastern Atacama. In: Reindel M, Wagner GA (eds) Natural science in archaeology. New technologies for archaeology: multidisciplinary investigations in Palpa and Nasca, Peru, Chapter 15. Springer, pp 245–270

Krbetschek MR, Degering D, Alexowsky W (2008) Infrarot-Radiofluoreszenz-Alter (IR-RF) unter-saalezeitlicher SedimenteMittel- und Ostdeutschlands. Z dt Ges Geowiss 159(1): 133–140

Kitis G, Liritzis I, Vafeiadou A (2002) Deconvolution of optical stimulated luminescence decay curves. J Radioanalytical Nuclear Chemistry 254:143–149

Lang A (1996) Die Infrarot-Stimulierte-Lumineszenz als Datierungsmethode für holozäne Lössderivate. – Heidelberger Geographische Arbeiten, Heidelberg, 103:137

Lang A, Hatté C, Rousseau D-D, Antoine P, Fontugne M, Zöller L, Hambach U (2003) High-resolution chronologies for loess: comparing AMS 14C and optical dating results. Quat Sci Rev 22(10–13):953–959

Li B, Li S-H (2011) Luminescence dating of K-feldspar from sediments: a protocol without anomalous fading correction. Quat Geochronol 6:468–479

Li S-H, Li B (2006a) Dose measurement using fast component of LM-OSL signals from quartz. Radiat Meas 41:534–541

Li B, Li S-H (2006b) Comparison of De estimates using the fast component and medium component of the quartz OSL. Radiat Meas 41:125–136

Li B, Li S-H, Wintle AG, Zhao H (2007) Isochron measurements of naturally irradiated K-feldspar grains. Radiat Meas 42:1315–1327

Li B, Li S-H, Wintle AG, Zhao H (2008a) Isochron dating of sediments using luminescence of K-feldspar grains. J Geophys Res 113:F02026. doi:10.1029/2007JF000900

Li B, Li SH, Wintle AG (2008b) Overcoming environmental dose rate changes in luminescence dating of water lain deposits. Geochronometria 30:33–40

Liritzis I (1994) A new dating method by thermoluminescence of carved megalithic stone building. C R l'Academie Sci, Paris, Serie II 319:603–610

Liritzis I, Galloway RB, Katsonopoulou D, Soters D (2001) In search of ancient Helike, Gulf of Corinth, Greece. J Coastal Res 17:(1):118–123

Liritzis I, Guilbert P, Foti F, Schvoerer M (1997) The temple of Apollo (Delphi) strengthens new thermoluminescence dating method. Geoarchaeol Int 12(5):479–496

Liritzis I, Galloway RB, Hong D, Kyparissi-Apostolika N (2002) OSL dating of three prehistoric ceramics from Theopetra cave, Greece: a case study. Mediterr Archaeol Archaeometry 2:35–43

Liritzis I, Zacharias N, Polymeris GS, Kitis G, Ernstson K, Sudhaus D, Neumair A, Mayer W, Rappenglück MA, Rappenglück B (2010a) The Chiemgau meteorite impact and tsunami event (southeast Germany): first OSL dating. Mediterr Archaeol Archaeometry 10(4):17–33

Liritzis I, Polymeris G, Zacharias N (2010b) Surface luminescence dating of 'Dragon Houses' and Armena Gate at Styra (Euboea, Greece). Mediterr Archaeol Archaeometry 10(3):65–81 Special Issue, (D. Keller, guest editor)

Mayya YS, Mortekai P, Murari MK, Singhvi AK (2006) Towards quantifying beta microdosimetric effects in single-grain quartz dose distribution. Radiation Measurements 41:1032–1039

Mejdahl V, Botter-Jensen L (1994) Luminescence dating of archaeological materials using a new techniques based on single aliquot measurements. Quat Sci Rev 13:551–554

Mejdahl V, Botter-Jensen L (1997) Experience with the SARA OSL method. Radiation Measurements 27:291–294

Murari MK, Achyuthan H, Singhvi AK (2007) Luminescence studies on the sediments laid down by the December 2004 tsunami event: Prospects for the dating of palaeo tsunamis and for the estimation of sediment fluxes. Current Sci 92:367–371

Murray AS, Wintle AG (2000) Luminescence dating of quartz using an improved single-aliquot regenerative-dose protocol. Radiation Measurements 32(1):57–73

Olley J, Caitcheon G, Murray A (1998) The distribution of apparent dose as determined by optically stimulated luminescence in small aliquots of fluvial quartz: Implications for dating young sediments. Quat Sci Rev 17:1033–1040

Olley JM, Caitcheon GG, Roberts RG (1999) The origin of dose distributions in fluvial sediments, and the prospect of dating single grains from fluvial deposits using optically stimulated luminescence. Radiation Measurements 30: 207–217

Porat N, Duller GAT, Roberts HM, Wintle AG (2009) A simplified SAR protocol for TT-OSL. Radiat Meas 44(5–6):538–542

Prescott JR, Huntley DJ, Hutton JT (1993) Estimation of equivalent dose in thermoluminescence dating – the Australian slide method: Ancient TL v. 11(1):1-5

Richter D, Krbetschek M (2006) A new thermoluminescence dating technique for heated flint. Archaeometry 48:695–715

Schilles T, Habermann J (2000) Radioluminescence dating: the IR emission of feldspar. Radiat Meas 32:679–683

Singarayer J, Bailey RM (2004) Component-resolved bleaching spectra of quartz optically stimulated luminescence: preliminary results and implications for dating. Radiat Meas 38:111–118

Singhvi AK, Williams MAJ, Rajaguru SN, Misra VN, Chawla S, Stokes S, Chauhan, N, Francis T, Ganjoo RK, Humphreys GS (2010) A ~200 ka record of climatic change and dune activity in the Thar Desert, India Quat Sci Rev, 29(23-24):3095–3105

Singhvi Ashok K, Stokes SC, Chauhan N (2011) Changes in natural OSL sensitivity during single aliquot regeneration procedure and their implications for equivalent dose determination. Geochronometria 38:231–241

Strahl J, Krbetschek MR, Luckert J, Machalett B, Meng S, Oches EA, Rappsilber I, Wansa S, Zöller L (2010) Geologie, Paläontologie und Geochronologie des Eem-Beckens Neumark-Nord 2 und Vergleich mit dem Becken Neumark-Nord 1 (Geiseltal, Sachsen-Anhalt). Eiszeitalter Gegenwart (Quat Sci J) 59(1–2):120–167

Trautmann T, Krbetschek MR, Dietrich A, Stolz W (1999a) Radioluminescence dating: a new tool for quaternary geology and archaeology. Naturwissenschaften 86:441–444

Trautmann T, Krbetschek MR, Dietrich A, Stolz W (1999b) Feldspar radioluminescence: a new dating tool and its physical background. J Luminescence 85:45–58

Trautmann T, Krbetschek MR, Stolz W (2000) A systematical study of radioluminescence properties of single feldspar grains. Radiat Meas 32:685–690

Vafiadou A, Murray AS, Liritzis I (2007) Optically stimulated luminescence (OSL) dating investigations of rock and underlying soil from three case studies. J Archaeol Sci 34:1659–1669

Vandenberghe DAG, Jain M, Murray AS (2009) Equivalent dose determination using a quartz isothermal TL signal. Radiation Measurements 44:439–444

Wang XL, Lu YC, Wintle AG (2006a) Recuperated OSL dating of fine-grained quartz in Chinese loess. Quat Geochronol 1:89–100

Wang W-D, Xia J-D, Zhou Z-X (2006b) Thermoluminescence dating of the ancient Chinese porcelain using a regression method of saturation exponential in pre-dose technique. Sci China Ser E: Technol Sci 49:194–209

Wang XL, Wintle AG, Lu YC (2007) Testing a single-aliquot protocol for recuperated OSL dating. – Radiation Measurements 42:380–391

Wagner GA, Krbetschek M, Degering D, Bahain J-J, Shao Q, Falgueres C, Voinchet P, Rightmire GP (2010) Radiometric dating of the type site for Homo heidelbergensis at Mauer, Germany. Proceedings of the National Academy of Sciences 107:19726–19730

Wintle AG, Huntley DJ (1980) Thermoluminescence dating of ocean sediments. Canadian J Earth Sci 17:348–360

Wintle AG, Murray AS (2006) A review of quartz optically stimulated luminescence characteristics and their relevance in single-aliquot regeneration dating protocols. Radiation Measurements 41:369–391

Yukihara EG, Mckeever SWS (2011) Optically stimulated luminescence: fundamentals and applications. Wiley, Sussex, p 362

Chapter 3
Dose Rate

The dose rate (D_R) can be expressed as

$$D_R = aD_\alpha + D_\beta + D_\gamma + D_{cos},$$

where α, β and γ indicate the contributions from individual components of nuclear decay of U, Th and K to the total dose. Cosmic ray dose depends on the location (latitude, longitude and altitude of the sample and can be computed using Prescott and Hutton (1994). Factor *a* is the alpha efficiency factor that accounts for lower luminescence induction efficiency of alpha particles compared to beta and gamma rays. The ranges of these different radiations in a solid medium like a mineral grain, varies considerably and hence the dose rate computation needs to account for grain size, range of radiation and their deposition inside and outside the sample. Normally, α- and β-dose are 'internal' to the artefact or sediment layer and γ- plus cosmic-dose constitute the 'external' dose or the environmental component. In nature, concentrations of the radioactive elements of ^{238}U, ^{235}U, ^{232}Th, ^{40}K and ^{87}Rb in the sample may differ from its surroundings. Thus, both the internal and external dose rates have to be assessed independently (Fig. 3.1).

The environmental dose is commonly measured by employing phosphors such as $CaSO_4$:Dy or Al_2O_3:C (Akselrod et al. 1998) that are encapsulated in copper vials of ~3 mm wall thickness (in consideration of radiation interaction effects) and buried at the sampling location for a period of up to several months (Kalchgruber et al. 2003). Such in situ measurements take into account possible seasonal variations in water content of the samples and heterogeneity of the sediment strata and may be assessed for β- and γ-components separateley (Kalchgruber and Wagner 2006). Normally heterogeneity within a radius of 30 cm is critical to dating application as this is the range of gamma rays. Where in situ measurements are not possible, and as a check on the field measurements, dose rate is additionally measured in the laboratory on the sample and any associated sediments, using appropriate geometry consideration for the irradiation environment arising from (often) complex heterogenous context.

I. Liritzis et al., *Luminescence Dating in Archaeology, Anthropology, and Geoarchaeology*, SpringerBriefs in Earth System Sciences,
DOI: 10.1007/978-3-319-00170-8_3, © The Author(s) 2013

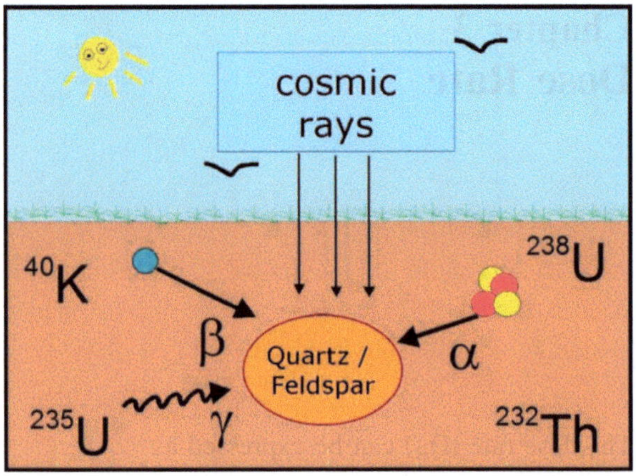

Fig. 3.1 Schematic illustration of all dose rate components involved in luminescence studies shown in the case of a quartz or feldspar grain

An important aspect is the estimation of dose-rate in the presence of water. While devoid of radioactivity itself, water absorbs radiation very differently than the sample and has the effect of 'diluting' the dose-rate to the sample by almost the same proportion as its amount. For a realistic estimation of total annual dose therefore, an estimation of water content is essential. The measured water content at the time of collection, the saturation capacity and the distance to ground water among others are used to arrive at an educated estimate of average water content during burial period.

Laboratory techniques employed for the estimation of U and Th isotope concentrations are, thick source ZnS (Ag) alpha-counting (Liritzis and Vafiadou 2012), beta-counting and low level gamma-spectrometry with the latter providing estimations for all isotopes involved. The choice of any of the above depends on the research facilities at hand, the site and the nature of the sample. Archaeological samples generally have limited available material, whereas larger samples are usually available in geology. Occasionally well calibrated portable γ-ray NaI scintillometers are used for onsite measurements (Liritzis and Galloway 1980). Aitken (1985) details the use of such gamma spectrometers and the use of stripping factors to account for the Compton background from higher energy gamma rays for each of the gamma ray energy windows. The measurement of environmental gamma dose rate needs careful attention to probe positioning, burial condition of a sample and radiation geometry of the adjacent context. Nevertheless, the NaI spectrometer is useful in inhomogeneous radiation fields that are often encountered in dating monuments and sediments.

Alpha-counting and alpha-spectrometric techniques are widely used for archaeological applications (Hossain et al. 2002; Sjostrand and Prescott 2002; Alpata et al. 2007) since the sample amount required is in the range of some grams in contrast to the tens of grams required for gamma spectroscopy measurements. These enable detection of disequilibrium (Michael et al. 2010a, b; Kokkoris and Liritzis 1997; NAA (used e.g. by Hilgers et al. 2001), and ICP-MS (e.g. Preusser and Kasper 2001).

Table 3.1 Updated dose rate data from Liritzis et al. (2012) including measurement errors and historical mean 1975–2011 (HM) along with standard deviation from the mean

Isotope	Alpha dose rate, Gy/ka	Beta dose rate, Gy/ka	Gamma dose rate, Gy/ka
U/ppm	2.797 ± 0.011	0.146 ± 0.0004	0.1118 ± 0.0002
HM	2.788 ± 0.0024	0.144 ± 0.006	0.118 ± 0.007
Th/ppm	0.7375 ± 0.0026	0.0275 ± 0.0009	0.0481 ± 0.0002
HM	0.736 ± 0.036	0.028 ± 0.001	0.048 ± 0.007
K 1 %	–	0.8011 ± 0.0073	0.2498 ± 0.0048
HM		0.816 ± 0.023	0.246 ± 0.003
Rb 50 ppm	–	0.0185 ± 0.0004	–
HM		0.021 ± 0.003	

The elemental concentration is converted into dose rate, using detailed conversion factors given by the recently updated conversion factors by Liritzis et al. (2012) (Table 3.1).

An important aspect of dose rate estimate is burial diagenesis due either to faunal or pedological issues resulting in the mixing of grains of different ages (Bateman et al. 2003). A linked observation is the early mineralogical studies of archaeological pottery samples that indicated significant variations in alkali metal concentrations due to environmental alterations during burial, affecting especially K concentration. Studies have established a relation between the calcareous nature of clay paste, the high-fired/over-fired state and the leaching of potassium. In Zacharias et al. (2007), a collection of archaeologically well dated Roman pottery was found to have experienced a leaching of ca. 50 % of their original K content. This provided underestimated dose rate values on average 18 %. Notably this study involved fine-grain TL dating, and if instead, coarse grain dating had been used, a significantly higher overestimation in age would have occured due to the (missing) higher contribution from potassium to the total dose rate.

References

Aitken MJ (1985) Thermoluminescence dating. Academic Press, London

Akselrod MS, Lucas AC, Polf JC, McKeever SWS (1998) Optically stimulated luminescence of Al2O3. Radiat Meas 29(3–4):391–399

Alpata B, Aisaa D, Bizzarria M, Blaskoa S, Espositoa G, Farnesinia L, Fioria F, Papia A, Postolachea V, Renzia F, Ionicab R, Manolescuc F, Ozkorucuklud S, Denizlie H, Tapanf I, Pilicerf E, Egidig F, Morettig C, Dicolag L (2007) Multipurpose high sensitivity radiation detector: terradex. Nucl Instrum Methods Phys Res A 57(4):479–492

Hilgers A, Murray AS, Schlaak N, Radtke U (2001) Comparison of quartz OSL protocols using Lateglacial and Holocene dune sands from Brandenburg, Germany. Quatern Sci Rev 20:731–736

Hilgers A, Murray AS, Schlaak N, Radtke U (2001) Comparison of quartz OSL protocols using Lateglacial and Holocene dune sands from Brandenburg, Germany. Quatern Sci Rev 20:731–736

Hossain SM, De Corte F, Vandenberghe D, Van den haute P (2002) A comparison of methods for the annual radiation dose determination in the luminescence dating of loess sediment. Nucl Instrum Methods Phys Res Sect A: Acceler Spectrometers Detectors Assoc Equip 490(3): 598–613
Kalchgruber R, Fuchs M, Murray AS, Wagner GA (2003) Evaluating dose-rate distributions in natural sediments using a-Al2O3:C grains. Rad Meas 37:293–297
Kalchgruber R Wagner GA (2006) Separate assessment of natural beta and gamma dose rates with TL from Al2O3:C single crystal chips. Rad Meas 41(2):154–162
Kokkoris M, Liritzis I (1997) Dose versus time for U-disequilibrium and revised dose-rate data for TL/ESR dating. Eur J PACT 45(IV.5):281–294
Liritzis Y, Galloway RB (1980) A new technique for calibrating a Nal (T I) scintillometer used to measure γ-dose rates in archaeological sites. Nucl Instrum Methods 174:593-597
Liritzis I, Vafiadou A (2012) Calibration aspects of thick source alpha counter ZnS system Measurement 45:1966–1980
Liritzis I, Stamoulis K, Papachristodoulou Ch, Ioannides KG (2012) Updating of radiation dose-rate conversion factors: a critical re-evaluation. Mediterr Archaeol Archa, Special Issue (Guest Editor: Dr C. Burbidge) 12(3):1–15
Michael CT, Zacharias N, Hein A (2010a) Accuracy testing using thick source alphaparticle spectroscopy for the U and Th series estimations. Appl Radiat Isot 68(4–5):858–860
Michael CT, Zacharias N, Hein A (2010b) Thick source alpha particle spectroscopy: possibilities and prospects. Mediterr Archaeol Archa 10(4):99–105
Prescott JR, Hutton JT (1994) Cosmic ray contributions to dose rates for luminescence and ESR dating: large depths and long-term time variations. Radiat Meas 23(2–3):497–500
Preusser F, Kasper HU (2001) Comparison of dose rate determination using high resolution gamma spectrometry and inductively coupled plasma-mass spectrometry. Ancient TL 19(1):19–23
Sjostrand H, Prescott JR (2002) Thick source alpha counting: the measurement of thorium. Ancient TL 20:7–10
Zacharias N, Schwedt A, Buxeda I, Garrigøs J, Michael CT, Mommsen H, Kilikoglou V (2007) A contribution to the study of post-depositional alterations of pottery using TL dating analysis. J Archaeol Sci 34:1804–1809

Chapter 4
Luminescence Dating of Archaeological Materials

This section deals with the kinds of chronological questions that luminescence can address in archaeology. Comparison with other chronometric methods in respect of resolving specific archaeological issues is discussed. The luminescence literature on archaeology is accessible in comprehensive bibliographies published in *Ancient TL* and is in the public domain (http://www.aber.ac.uk/ancient-tl/). Reviews by Roberts (1997), Liritzis (2000), Feathers (2003), Richter (2007) describe earlier developments.

Every event being dated has a duration. For discrete artifacts like ceramics or lithics, the events being dated—the manufacturing period or discarding—are close in time, so that these can not be resolved (being within the typical standard error of the measured ages), and only single age estimations are possible. For occupations, durations can be longer extending even up to several centuries. In such cases single age estimates are then no longer sufficient and multiple age determinations are necessary to estimate the time range (e.g., Lipo et al. 2005).

Luminescence dating has not realized its potential in archaeological applications. Since 1990s work on discrete artifacts, pottery and lithics, has been supplemented by the dating of large, complex artifacts, namely the architecture of monuments, where the duration of manufacture can be long. While sometimes the age of construction is not known at all, often some age information is known and luminescence dating has been applied to determine the construction sequence and duration.

4.1 Pottery and Other Fired Ware

An often remarked benefit of luminescence dating when applied particularly to pottery is the capability of direct dating i.e. the event dated is the same as the event of interest (i.e., manufacture of the pottery), eliminating the need for bridging arguments to associate the two. The need for differentiation between the dating event, the target event and the bridging arguments that are necessary to link the

I. Liritzis et al., *Luminescence Dating in Archaeology, Anthropology, and Geoarchaeology*, SpringerBriefs in Earth System Sciences,
DOI: 10.1007/978-3-319-00170-8_4,

two was first clearly stated by Dean (1978) in the context of tree-ring dating and has since been explored in luminescence dating of lithics (Richter 2007; Richter et al. 2009) and pottery (Dykeman et al. 2002; Yang et al. 2005; Benea et al. 2007; Feathers 2009).

Besides pottery other dateable objects are vases, bricks, tiles, porcelains and other earthen materials that have been heated to at least 500 °C. In all these the basic signal arises from quartz and/or feldspars. By heating, these materials undergo physicochemical alterations of the initial clay components and their admixtures/additives, towards creation of ceramic phases and at the same time erase the geological luminescence of minerals contained in pottery to reset the 'luminescence clock'. Exception are the adobe mud-made sun-dried clay objects where the bleaching occurs from sunlight.

In the case of *pottery,* the most recent heating is assumed to represent the manufacturing event, or for pottery that is used with high heat, such as cooking ware, a use event, soon after manufacturing. If the manufacturing event is what the archaeologists want to know, then luminescence methods provide a direct date, provided any later date of firing resetting the luminescence can be ruled out. However, often the primary archaeological interest is not the manufacturing event, but is the depositional event, an event often called an, 'occupation' that represents the period of time when a certain group of artifacts came together in restricted space. That the manufacturing event and depositional date might differ significantly in age is often ignored but this aspect can be non trivial. Thus for example, heirlooms from an earlier era and intrusions from a later period are not uncommon, and differences in time between manufacturing and deposition may span hundreds of years (e.g., Holt and Feathers 2003; Fitzpatrick et al. 2009). Dating several samples of pottery from a single stratum has shown that often these are not of the same age (Feathers 2008, 2009). The potential for luminescence to determine a distribution of ages for pottery sherds from a single site, analogous to the distribution of ages from single grains in a sediment sample to identify mixing, has largely not been explored (see Barnett 2000), partly because of costs and the effort needed for the analyses. Yet, such study can make archaeological interpretations more precise and lead to better understanding of cultural change. However, the different cultural periods in question must span time intervals which can be resolved with the typical error margins of luminescence dating.

Methodological developments in dating pottery have progressed only a little beyond the original protocols of the 1980s (Aitken 1985). An important development has been the use of OSL for dating pottery (Liritzis et al. 1994, 1997a; Aitken 1998; Thomas et al. 2008; Benea et al. 2007), on the premise that the heating event also resets the optically sensitive traps. This is a welcome development as routine dating of heated/fired materials using the SAR-OSL procedure is more practicable when the sample availability of datable material is limited and replicate measurements are needed.

A perfected and automated OSL method for the dating of pottery could be the way for rapid and cost-effective throughput of samples. OSL has been applied both to coarse quartz inclusions and to polymineral fine-grains (both are available in pottery), the latter using double SAR protocol where the IR stimulation precedes the blue stimulation. The IR stimulation removes the potential interference of feldspar grains to the Blue light stimulated luminescence (Jain and Singhvi 2001; Banerjee et al. 2001). For such a post-IR, blue signal (post-IR BLSL) for fine-grains may be free of anomalous fading, given that the IRSL signals from fine-grained pottery extracts are often quite weak (Feathers 2009). It is however prudent to measure the extent of fading and apply appropriate corrections to the post-IR signal (either post-IR BLSL or post-IR IRSL) that may still be required (Lamothe 2004). Fading corrections on polymineral fine-grained material have provided successful results that accord with independent controls (Lamothe 2004; Feathers 2009; Sánchez et al. 2008), The use of post-IR IRSL may also herald newer possibilities for the dating of pottery which have higher contribution of signal from feldspars. Experience, however, suggests that the firing event increase the sensitivity of quartz and decreases that of feldspars, so in the net the quartz OSL generally dominates.

For young ceramic samples, pre-dose dating, which utilizes the 110 °C quartz peak, may be useful and may also circumvent fading because it comes exlusively from quartz (Galli et al. 2006; Park et al. 2005; Wang et al. 2006; Zimmerman 1971). Pre-dose methods may also be appropriate for very small and young samples as the predose sensitization saturates at about 10 Gy. Pre-dose dating relies on the fact that the sensitivity of 110 °C depends on the past irradiation and thermal history and with appropriate laboratory calibration it provides an accurate measure of paleodoses. Given that sensitivity of the 110 °C peak of quartz is significant (in response to annealing and irradiation) (Liritzis 1980, 1982; Sunta and David 1982) accurate measurements are possible.

In this the major effort is to ensure that the applied laboratory protocols sufficiently correct for sensitivity changes due to thermal annealing and the known pre-dose effect; (Zimmerman 1971; Rendell et al. 1994; Vieillevigne et al. 2006). The fast component OSL signals also shows similar effects of fading and sensitivity changes (Koul and Chougaonkar 2007).

Work with TL on pottery has been reported for many parts of the world and includes both excavated finds and museum objects, although with the latter precision will be less because of an unknown external dose rate (e.g. Cosma et al. 2006; Yang et al. 2005; Liritzis and McKerrell 1979; Galli et al. 2004; Feathers 2008, 2009; Barnett 2000; Chiavari et al. 2001; Feathers 2000; Guibert et al. 1994, 2001; Herbert et al. 2002; Sampson et al. 1972, 1997; Whittle and Arnaud 1975). Glassy nature of porcelain has required special methods (Wang 2008). The use of the red emission band in TL for dating pottery, claiming less problem with anomalous fading, has also been attempted, (Hashimoto et al. 2005; Nakata et al. 2007).

A good example of application of luminescence dating to pottery is from India. Hot and humid weather in tropical regions implies poor preservation of radiocarbon dateable organic material. This has paved the way for luminescence dating in India and it was initially meant to supplement the radiocarbon based Indian archaeology. Initial application was on several pottery bearing sites including the Harappan and the pre-Harappan and the grey ceramic wares. An important new work then was the dating of archeological sites mentioned in the India Epic Ramayana (Agarwal et al. 1981). This work was based on the observation that the sites mentioned in the epic bear a reasonable resemblance in terms of distances and other geographic features to archaeological sites. It was argued that the epic would have been initiated when similar cultural materials appeared simultaneously at all the sites. The black slipped ware was the earliest common culture that existed at all the sites and was dated by luminescence to 2700 years. Thus archeologically, the antiquity of the epic could not go beyond 2700 years. The dating of a pottery megalithic burial pit was placed at 2900–3400 years. Nambi and Murty (1981) dated an upper paleolithic hearth to 17.39 ± 1.74 ka which was much younger that a radiocarbon age of 25 ka ^{14}C BP, suggesting that the radiocarbon age was overestimating. Luminescence ages on pottery were also used to establish the dating of desert sands (Singhvi et al. 1982). Chawla and Singhvi (1989) were first to demonstrate that the luminescence age of archeological sediment and the pottery in it was the same, thereby paving the way for the direct dating of archaeological horizons, treating them as ordinary sediment. This application was based on the fact that the net accretion rate at any archeological site is low and the human and biological activity results in constant churning of the surface sediment. This results in onsite bleaching of luminescence of minerals that constitute the archeological strata. Other pottery based archeological work in India has been summarized by Singhvi et al. (1991).

Luminescence in pottery has other applications than dating. Sunta and David (1982), as well as, Polymeris et al. (2006) used the change in sensitivity of TL and OSL signals to detect original firing temperatures. Luminescence intensities (Zacharias et al. 2008), TL and OSL decay curves (Liritzis 1997; Liritzis et al. 2008), and the emission spectrum from radioluminescence (Galli et al. 2004) have been used to characterize mineralogy. Beta and gamma emissions from ^{40}K provide a large portion of the dose rate, and where leaching has occurred, ages can be severely overestimated, depending on when the leaching occurred. Zacharias et al. (2007) measured current dose rates and equivalent dose on known age sherds to estimate the time of leaching of other sherds from the same site. Leaching of carbonate temper changes the internal dose rate as discussed by Feathers (2009).

4.2 Lithics

Most current research in TL involves *lithics*, particularly microcrystalline forms such as *flints* and *cherts*, where OSL does not appear to work (Richter and Temming 2006). Other kinds of rocks that were used prehistorically such as fire-altered stones from hearths, may also be perfectly amenable to OSL/TL dating, as long as they were fired to sufficiently high temperatures (Rhodes et al. 2009). One recent application on chipped stone was OSL dating of lithic microdebitage from quartzites, which can be distinguished from quartz sand particles morphologically using scanning electron microscopy (Susino 2010). This application assumes resetting of the signal by daylight and is thus a depositional event. Like for other artifacts, and if post-depositional reworking may be ruled out, it may be considered direct dating of a manufacturing event.

For the dating of lithic implements, the resetting mechanism is primarily heat. At the time of their manufacture, often lithics were heat-treated to improve their flaking properties. In the process the geological luminescence was reset, providing a way to date the manufacturing event. Often, however, such heat-treatment did not go to sufficiently high temperatures, not more than around 300–400 °C (Dunnell and Feathers 1995). More commonly, the dating event corresponds to chipped stone artifacts being deposited into a hearth and fired, and therefore reflects occupations, or the event when the fire was used, rather than their manufactures. Richter et al. (2009) argue that dating of lithic implements is to be preferred to other types of materials to establish the chronometry of occupations of Palaeolithic sites. OSL dating of the enclosing sediments is less reliable because of a greater likelihood of post-depositional processes that could move sediment grains around (e.g., Richter 2007).

At many sites, dating the lithics or ceramics are the *only* chronometric dating options, because bridging arguments to connect events dated by other methods to human activity are too convoluted to be credible. This is particularly true of surface finds, where e.g. the chronological association with any charcoal, if at all preserved, is tenuous. Dating of ceramic finds on the current surface is not widespread, partly because of concern with uncertainties surrounding the external dose-rate, including that stemming from radon fallout, while any reported TL ages during and post Chernobyl fallout should be cautiously considered if not subjected a possible correction (Liritzis 1987). But these uncertainties are probably less than those surrounding burial, as shown by Dunnell and Feathers (1995). All artifacts, after all, were once on the surface for some, often unknown period of time. Aside from some work with *hearth rocks* (Rhodes et al. 2009), little work has been done toward direct dating of lithic tools from surface contexts. This is a potentially important application that merits further research. Dating flints or cherts may be more problematic in this regard than ceramics, because flints or cherts have low internal dose rates to offset the uncertainties in the external dose rates. The dose rate of surface calcitic lithics may be dominated by the cosmic ray and/or environmental ground gamma dose.

Even at buried sites, the ubiquitous occurrence of ceramics or lithics dateable by luminescence still remains the best dating option. Though radiocarbon may offer better measurement precision, uncertainty in bridging arguments connecting the radiocarbon dated material to the archaeological events limits its accuracy. In addition, during some periods of time such as during the past millennium when the radiocarbon calibration curve is relatively flat, luminescence dating may provide a higher precision (e.g., Feathers 2008). Also the dating range of luminescence is an order of magnitude higher than that of radiocarbon. (e.g., Jennings et al. 2009) making it possible to date early stone age cultures. Application of TL dating to lithics at Paleolithic sites, has provided chronological information that is unavailable otherwise. A better understanding of the spread of anatomically modern humans and their replacements of more archaic forms such as Neanderthals has been made possible by luminescence. Such work, sometimes combined with OSL of associated sediments, continues apace in Europe, the Middle East and Africa (Singhvi et al. 1986a, b, c; 1989; Tribolo et al. 2009; Jennings et al. 2009; Adler et al. 2008; Richter et al. 2008; Barton et al. 2008; Bouzouggar et al. 2007; Valladas et al. 2005, 2007, 2008; Mercier et al. 2007a, b; Richter 2007; Tribolo et al. 2006).

A limiting factor in applying luminescence dating to lithics is the sample amount. The most reliable method for dating flints or cherts is a MAAD method using blue light emission, but this usually requires a samples size of $\sim$10 g (Richter and Krbetschek 2006). This restricts applications to a few archaeological sites. Richter and Krbetschek (2006), as well as, Richter and Temming (2006), used an abbreviated SAR protocol measuring the orange-red emission with good results from small samples. Further exploration into dating small lithic samples is needed.

Accurate estimate of the external dose rate is a key to the precision of produced ages. Ceramics are less problematic, because being made of clay; they generally have a higher internal dose rate, generally thrice the external dose rate. They are also well mixed during manufacture, resulting in a more homogeneous internal dose rate.

Flints or cherts have low to negligible internal dose rates. This implies that the precision and accuracy of the external dose rate determines the precision and accuracy of the ages. For the cave site of Hayoinim in Israel, Mercier et al. (2007a, b) demonstrated that accurate results can be obtained with in situ dosimetric measurements. Use of 76 dosimeters provided a three-dimensional view of the dose rate across the site. Further, different dose rates could be correlated with mineralogy. This work facilitated dating of over 70 lithic implements by plotting the dose rates on a dose-rate map. As expected, samples close to mineralogical boundaries or associated with sediments undergoing U uptake resulted in less precise dates. An additional problem in the dating of lithics is the heterogeneity in the distribution of radionuclides, and in particular the heterogeneity in the distribution of the alpha dose rate. Tribolo et al. (2006) measured such heterogeneity with fission track and other methods.

TL dating of burnt flint/chert artifacts has been the main chronometric method used for prehistoric archaeological contexts covering the time span of the controlled use of fire by humans. TL dates on burnt flint artefacts from Lower and Middle Palaeolithic of the Near East, are widely used as a reference in debates on the evolution of Palaeolithic industries and on the origin of modern humans and their relationship to the Neanderthals (Mercier and Valladas 2003). A detailed TL study (Valladas et al. 2007) on burnt flints from the lowest Middle Palaeolithic stratigraphic unit of Theopetra cave at Thessaly (Greece), verified that these layers are much older than was postulated on the basis of earlier radiocarbon dates and also that Theopetra contains the oldest, dated lithic artefacts of the Greek Middle Palaeolithic, so far.

4.3 Luminescence Dating of Surfaces

The chronometric age determination of lithic artefacts and stone structures (tools, monoliths, buildings, cairns, field walls etc.) is usually achieved by dating associated materials, such as radiocarbon (C-14) datable remains. However, in many cases appropriate debris is either not available, or the association with the archaeology is insecure. In such case luminescence dating may be the solution, not by dating the lithic material itself but dating the moment when its surface was the last time exposed to light.

During the process of working the stone blocks (cutting and carving, or sculpturing) the solar day-light bleaches the previously stored luminescence in the surface, down to a depth determined by the depth of penetration of light in that material (Fig. 4.1). This exposure—just minutes required for granite, basalt or sandstone—erases the luminescene to a zero or a near zero residual value. On construction or sedimentary covering, the rock surface is shielded from light so that the luminescene signal reaccumulates again due to irradiation from ambient radioactivity. The signal growth continues until excavation and measurement. The equivalent dose and the dose rates then provide the age of burial of the surface i.e. the construction of the dwelling.

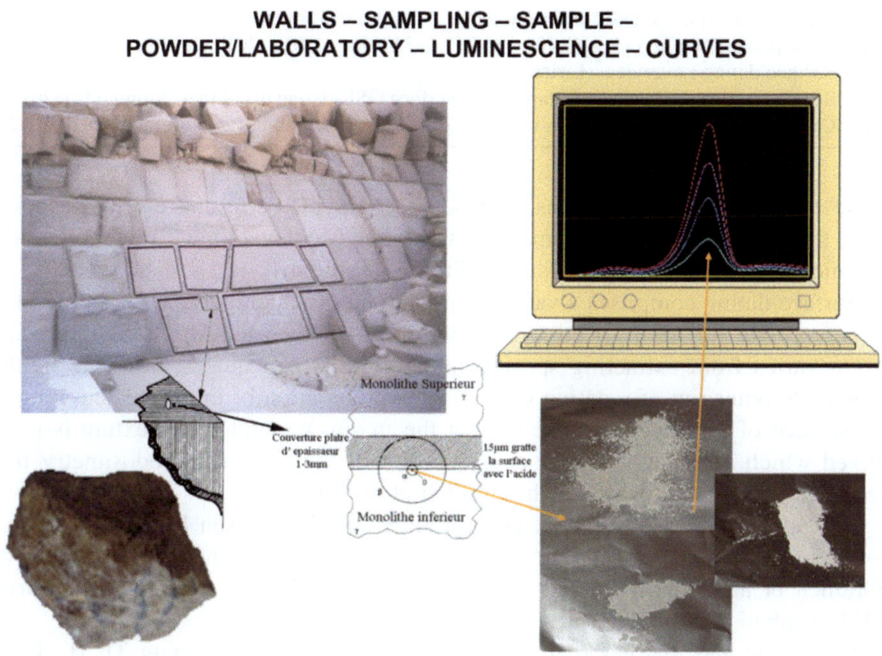

Fig. 4.1 Sampling at ancient monuments. An example of Mykerinus pyramid, Giza. (Liritzis 2011)

Table 4.1 Chronological development of novel luminescence rock surface dating protocols and applications (including relevant instrumentation progress)

1. Exposure to daylight of sediments (Wintle and Huntley 1980).
2. Thermoluminescence response of archaeological stone sculpture (Vaz 1983)
3. Exposure of rock surfaces to daylight—dating of monuments (Liritzis 1994)
4. Dating of Quartzite pebbles (Huntley and Richards 1997)
5. TL dating of temple of Apollo Delphi (Liritzis et al. 1997d)
6. TL dating of two Pyramidals (Theocaris et al. 1997)
7. Spatially resolved OSL with charge-coupled devises and confocal microscopy (Duller et al. 1997)
8. Dating of marble monuments and objects (Liritzis and Galloway 1999)
9. Surface dating - methodological aspects (Habermann et al. 2000)
10. High resolution detection technique (HR-OSL) (Greilich et al. 2002)
11. OSL on volcanic lithic artefacts (Morganstein et al. 2003)
12. Surface dating of monuments (Liritzis and Vafiadou 2005)
13. Rock surface dating of geoglyphs from Peru and castle in Germany (Greilich et al. 2005)
14. Dating of Medieval usage of the Externstein Cave in Germany by OSL-analysis of heated regions from the cave wall by C. Woda (in Kadereit et al. 2007)
15. Dating of Granitic cobble overlying sediment floor (Vafiadou et al. 2007)
16. Petrologic investigations of Egyptian limestone monuments (Liritzis et al. 2007)
17. OSL and TL properties of carved rock types (Liritzis et al. 2008)
18. HR-OSL dating of granitic cobble surface from Huyaco deposits in southern Peru (HR-OSL dating) (Kadereit et al., 2009)
19. OSL dating of quartz from calcitic cyclopean blocks ("dragon houses"- Euboea, Greece, (Liritzis et al. 2010a)
20. OSL investigations of coastal rock surfaces (Sohbati et al. 2011)
21. Quartz technique for OSL of limestones and quartzite (Liritzis et al. 2010b)
22. Solar penetration in rock surfaces (Liritzis and Laskaris 2011, 2010b, c)
23. Surface dating : overview (Liritzis 2011)
24. Reconstructing sea levels in Antarctica by surface OSL-dating of cobbles (Simms et al. 2011)
25. OSL dating of quartzite cobbles from the Tapada do Montinh Portugal (Sohbati et al. 2012)

Several application examples from Greece, Egypt, Peru, and elsewhere covering the period third millennium to Classical and Medieval times, are reported. A chronicle of these developments is outlined in Table 4.1.

Surface dating comprises a variety of techniques including TL, OSL and IRSL, as appropriate for the respective material and dating event. Also sample preparation varies from extracting specific minerals, like e.g. quartz, from the light-penetrable outer rim of a dating object to the preparation of thin slices cut out of the surface of a respective stone so that the in situ mineralogical texture is preserved which potentially allows consideration of the in situ microdosimetry for dating (HR-OSL dating approach; Greilich et al. 2002).

In dealing with dating of ancient buildings and buried sun bleached objects/ cobbles careful sampling and processing is needed and these have been discussed (Greilich et al. 2005; Vafiadou et al. 2007; Liritzis 2010; Liritzis and Laskaris 2011; Sohbati et al. 2011).

Limestone and marbles (calcite, $CaCO3$) have been dated using TL (Liritzis 2001), and granites, basalt and sandstone have been dated using either OSL or TL, with better results from OSL. Measuring De in calcites using OSL has not been successful (Galloway 2002). Instead quartz extracted from limestone surfaces has been proposed (Liritzis et al. 2007, 2010b).

Error Evaluation of Surface Dating

There are various sources of error:

(1) For MAAD scatter arising from differences in the radiation response of different grains of a same mineral inspite of normalization procedure. This leads to scattering of additive dose points, which to some extent can be compensated by a large number of replicate measurements.
(2) Errors in the measurement of low environmental radiation dose rates, particularly the gamma ray contribution. Use of multiple methods is recommended, and care in the counting geometry, accounting also for possible sand/soil cover during the past and that the water content is warranted.
(3) Destruction of surface datable layer due to friction, weathering and erosion, development of salts and secondary minerals, and moss/lichens. Meticulous examination and handling is needed to remove secondary surface effects. A safer sampling procedure is to divide the inner block surface into several sub areas and this way a geological D_e obviously derived from unidentified drifting (friction) is easily recognized as outlier and excluded (Liritzis et al. 2010a, b).
(4) Incomplete bleaching. If C-14 dating is available and shows significantly younger ages, this may point to insufficient bleaching, a situation that would suggests that the rock was not completely bleached, a situation that would result in a bi- or multi-modal distribution of equivalent doses. There can never be a guarantee that stones used to make burial mounds were adequately exposed to light before final burial (Liritzis et al. 1997d) and hence notionally, the OSL ages should serve as upper bounds. At any rate the residual luminescence from incomplete bleaching may be identified from dose-temperature plateau tests (Liritzis et al. 1997d; Liritzis 2010).

The OSL dating errors on megalithics may range between ±5–7 %, while the TL dates on limestones carry an error of around ±7–20 %. The latter errors can improve to ±5–7 % by using the quartz inclusion technique.

4.4 Other Artifacts (bones, mortars, bricks, mounds)

Bone tools can possibly be dated by luminescence. Two possible events can be dated: the time of bone crystallization or the heating event of bone. Early attempts using TL were not successful, but recent work on the use of IRSL to bone and teeth, yielded signals that responded to radiation dose. The signals were of low sensitivity but were bleachable by light (Meric et al. 2008). The signal arises from an inorganic fraction and hence the organic fraction has to be removed. Additional research on the reliable use of bone for dating is awaited.

Applications have taken advantage of resetting by heat, such as for *bricks* (Blain et al. 2007, 2009; Vieillevigne et al. 2007; Chruscinska et al. 2008; Bailiff 2007), and resetting by light, such as for *brick or stone surfaces* (Vieillevigne et al. 2006; Liritzis et al. 2010a, b), for *mortar* (Feathers et al. 2008), for earthen structures such as *mounds* (Bush and Feathers 2003), for *canals* (Sanderson et al. 2003, 2007, Berger et al. 2009), and for *agricultural terraces* (Borejsza et al. 2008).

Lime mortars mixed with sand were extensively used in architectural and decorative works during the last 4,000 years. The first application of SAR-OSL to quartz extracted from mortars referred to a Byzantine church monument dated to the tenth century (Zacharias et al. 2002); the same premises can be applied to any cemented or calcite material containing quartz (Sect. 4.3).

Issues of partial bleaching for poorly mixed *mortar* have been discussed by Feathers et al. (2008) and, for water-lain canal sediments by Sanderson et al. (2007). Most of these studies, particularly those dealing with bricks/stones and mortar, have also had to deal with complicated external dose rates.

4.5 Slags, Pyrometallurgical Remains

Ancient slags present a valuable source to study the technology of early metal production. Crucial to the understanding of these sites is their dating, often complicated by the rarity of contextual diagnostic pottery fragments necessitating the use of appropriate scientific techniques. Luminescence has a great potential for absolute dating of archaeometallurgical remains, including slags, kiln fragments, fired or vitrified linings of furnaces, hearths, tuyères, crucibles etc. (Godfrey-Smith and Casey 2003) due to the resetting of the 'luminescence clock' caused by the temperature of the smelting process. Direct dating of slag by means of luminescence has been problematic so far both with TL (Elitzsch et al. 1983) and OSL (Gautier 2001) due to the complex composition of the material and inaccuracies in microdosimetry calculations. In Zacharias et al. (2006a, b), TL dating of archaeometallurgical kiln wall fragments from two Cycladic islands (Kythnos and Seriphos) is reported. TL dating of selected layers from the fragments indicate that the metallurgical activities for both islands took place in the first half of the third millennium BC, i.e. during the Aegean Early Bronze Age periods. The red TL of quartz from archaeometallurgical slag was exploited by Haustein and Krbetschek 2002; Haustein et al. 2003.

References

Adler DS, Bar-Yosef O, Belfer-Cohen A, Tushabramishvili N, Boaretto E, Mercier N, Valladas H, Rink WJ (2008) Dating the demise: Neandertal extinction and the establishment of modern humans in the southern Caucasus. J Hum Evol 35:817–833

Agrawal DP, Bhandari N, Lal BB, Singhvi AK (1981) Thermoluminescence Dating of Pottery from Sringaverapura: A Ramayana Site. Proc Ind Acad Sci 90:E161

Aitken MJ (1985) Thermoluminescence Dating. Academic Press, London
Aitken MJ (1998) An introduction to optical dating. Oxford University Press, Oxford. 280p
Bailiff IK (2007) Methodological developments in the luminescence dating of brick from English Late-Medieval and Post-Medieval buildings. Archaeometry 49(4):827–851
Banerjee D, Murray AS, Bøtter-Jensen L, Lang A (2001) Equivalent dose estimation using a single aliquot of polymineral fine grains. Radiation Measurements 33:73–94
Barnett SM (2000) Luminescence dating pottery from later prehistoric Britain. Archaeometry 42:431–457
Barton RNE, Bouzouggar A, Humphrey LT, Berridge P, Collcutt SN, Gale R, Parfitt S, Parker AG, Rhodes EJ, Schwenninger JL (2008) Human burial evidence from Hattab II Cave (Oued Laou-Tétuoan, Morocco) and the question of continuity in Late Pleistocene- Holocene mortuary practices in Northwest Africa. Cambridge Archaeological J 18(2):195–214
Benea V, Vandenberghe D, Timar A, van den Haute P, Cosma C, Gligor M, Florescu C (2007) Luminescence dating of Neolithic ceramics from lumea noua, Romania. Geochronometria 28:9–16
Berger GW, Post S, Wenker C (2009) Single and Multigrain Quartz-Luminescence Dating of Irrigation-Channel Features in Santa Fe, New Mexico. Geoarchaeology 24(4):383–401
Blain S, Guibert P, Bouvier A, Vieillevigne E, Bechtel F, Sapin C, Baylé M (2007) TL-dating applied to building archaeology: The case of the medieval church Notre-Dame-Sous-Terre (Mont-Saint-Michel, France). Radiat Meas 42:1483–1491
Borejsza A, Rodriguez Lopez I, Frederick CD, Bateman MD (2008) Agricultural slope management and soil erosion at La Laguna, Tlaxcala, Mexico. J Archaeological Science 35:1854–1866
Bouzouggar A, Barton N, Vanhaeren M, d'Errico F, Collcutt S, Higham T, Hodge E, Parfitt S, Rhodes E, Schwenninger J-L, Stringer C, Turner E, Ward S (2007) 82,000-year-old shell beads from North Africa and implications for the origins of modern human behavior. Proc Nat Acad Sci 104:9964–9969
Bush DA, Feathers JK (2003) Application of OSL single-aliquot and singlegrain dating to quartz from anthropogenic soil profiles in the SE United States. Quaternary Science Reviews 22:1153–1159
Chawla S, Singhvi AK (1989) Thermoluminescence dating of archaeological sediments. Naturwissenschaften 76(9):416
Chiavari C, Martini M, Sibilia E, Vandini M (2001) Thermoluminescence (TL) characterization and dating feasibility of ancient glass mosaic. Quaternary Science Reviews 20:967–972
Chruscinska A, Jesionowski B, Oczkowski HL, Przegiekla KR (2008) Using the TL single-aliquot regenerative dose protocol for the verification of the chronology of the Teutonic Order Castle in Malbork. Geochronometria 30:61–67
Cosma C, Benea V, Timar A, Bardos D, Paunoiu C (2006) Preliminary dating results for ancient ceramics from Romania by means of thermoluminescence. Radiat Meas 41:987–990
Dean JS (1978) Independent dating in archaeological analysis. Advances in Archaeological Method and Theory 1:223–255
Duller GAT, Botter-Jensen L, Markey BG (1997) A luminescence imaging system based on a CCD camera. Radiation Measurements 27:91–99
Dunnell RC, Feathers JK (1995) Thermoluminescence dating of surficial archaeological material. In Dating in Surface Context (edited by C. Beck). University of New Mexico Press, Alburquerque, pp. 115–137
Dykeman DD, Towner RH, Feathers JK (2002) Tree-ring and thermoluminescence dating: evaluating methods for confidently dating protohistoric Navajo sites. American Antiquity 67:145–164
Elitzsch E, Pernicka E, Wagner GA (1983) Thermoluminescence dating of archaeometallurgical slags. PACT 9:271–286
Feathers JK (2000) Why luminescence dating deserves wider application in American archaeology. In, It's About Time, edited by Stephen Nash. University of Utah Press, Salt Lake City, pp. 152–166

Feathers JK (2003) Use of luminescence dating in archaeology. Meas Sci Technol 14:1493–1509
Feathers JK (2008) Absolute dating in the Mississippi Delta, In Rafferty, J., and Peacock, E., (eds.) Time's River: Archaeological Syntheses in the Yazoo Basin and Lower Mississippi River Valley. The University of Alabama Press, Tuscaloosa, pp. 168–181
Feathers JK, Johnson J, Kembel S (2008) Luminescence Dating of Monumental Architecture at Chavín da Huantár, Peru. J Archaeological Method and Theory 15:266–296
Feathers JK (2009) Problems of ceramic chronology in the Southeast: Does shelltempered pottery appear earlier than we think? American Antiquity 74:113–142
Fitzpatrick SM, Kaye Q, Feathers JK, Pavia JA, Marsaglia KM (2009) Evidence for inter-island transport of heirlooms: luminescence dating and petrographic analysis of ceramic inhaling bowls from Carriacou, West Indies. J Archaeological Science 36:595–606
Galloway RB (2002) Luminescence lifetimes in quartz: dependence on annealing temperature prior to beta irradiation. Radiation Measurements 35:67–77
Galli A, Martini M, Sibilia E, Padeletti G, Fermo P (2004) Luminescence properties of luster decorated majolica. Appl Phys A: Mater Sci Process 79:293–297
Galli A, Martini M, Montanari C, Panzeri L, Sibilia E (2006) TL of fine-grain samples from quartz-rich archaeological ceramics: Dosimetry using the 110 and 210°C TL peaks. Radiat Meas 41:1009–1014
Gautier A (2001) Luminescence dating of archaeometallurgical slag: use of the SAR technique for determination of the burial dose. Quatern Sci Rev 20:973–980
Godfrey-Smith DI, Casey JL (2003) Direct thermoluminescence chronology for early iron age smelting technology on the Gambaga Escarpment, Ghana. J Archaeol Sci 30:1037–1050
Greilich S, Glasmacher UA, Wagner GA (2002) Spatially resolved detection of luminescence: a unique tool for archaeochronometry. Naturwissenschaften 89:371–375
Greilich S, Glasmacher GA, Wagner GA (2005) Optical dating of granitic stone surfaces. Archaeometry 47(3):645–665
Guibert P, Ney C, Bechtel F, Shvoerer M, Geus F (1994) TL and radiocarbon dating of Neolithic sepultures from Sudan: intercomparison of results. Radiation Measurements 23:393–398
Guibert P, Vartanian E, Roque C, Bechtel F, Schvoerer M (2001) Luminescence dating of burnt materials: effects of preheat treatment on OSL and consequences for dating procedures. Radiation Measurements 33:439–444
Habermann J, Schilles T, Kalchgruber R, Wagner GA (2000) Steps towards surface dating using luminescence. Radiation Measurements 32(5–6):847–851
Hashimoto T, Yawata T, Takano M (2005) Comparison of naturally accumulated radiation-doses between RTL, BTL, OSL, and IRSL using white minerals from burnt archaeological materials and usefulness of RTL-dating from quartz extracts. Geochem J 39:201–212
Haustein M, Krbetschek MR (2002) The red thermoluminescence of quartz and its application in dating archaeometallurgical slags. Radiation Protection Dosimetry 100:375–378
Haustein M, Roewer G, Krbetschek MR, Pernicka E (2003) Dating archaeomettalurgical slags using thermoluminescence. Archaeometry 45(3):519–530
Herbert JM, Feathers JK, Cordell AS (2002) Building ceramic chronologies with thermoluminescence dating: a case study from the Carolina Sandhills, Southeastern Archaeology 21:92–108
Holt JZ, Feathers JK (2003) Dating the Middle to Late Woodland transition in the Illinois valley: radiocarbon and thermoluminescence dates from the Baehr-Gust site, Midcontinental. J Archaeology 28:73–93
Huntley DJ, Richards MP (1997) The age of the Diring Yuriakh archaeological site. Ancient TL 15:2–3
Jain M, Singhvi AK (2001) Limits to depletion of green light stimulated luminescence in feldspars: implication for Quartz dating. Rad Meas 33:883–892
Jennings RP, Pacheco FG, Barton RNE, Collcutt SN, Gale R, Gleed-Owen CP, Gutiérrez López JM, Higham TFG, Parker A, Price C, Rhodesm E, Santiago Pérez A, Schwenninger J-L, Turner E (2009) New dates and palaeoenvironmental evidence for the Middle to Upper

Palaeolithic occupation of Higueral de Valleja Cave, southern Spain. Quatern Sci Rev 28:830–839

Kadereit A, Wagner GA, Corvinus G (2007) OSL/IRSL fine-grain dating. A preliminary chronology of Quaternary slopewash and alluvial deposits from the Dun valleys of Dang and Deokhuri in western Nepal. In: Freund G, Reisch L (Hrsg.) (eds) Prehistoric cultures in Nepal from the early palaeolithic to the neolithic and the quaternary geology of the Dang-Deokhuri Dun Valleys, Vol. I, Appendix II. Harrassowitz Verlag, Wiesbaden, pp 372–379

Kadereit A, Greilich S, Woda S, Wagner GA (2009) Cold light from the sediments of a hot desert: how luminescence dating sheds light on the landscape development of the north-eastern Atacama. In: Reindel M, Wagner GA (Hrsg.) (eds) Natural science in archaeology. New technologies for archaeology. Multidisciplinary investigations in Palpa and Nasca, Peru, Chapter 15: 245–270, Springer

Koul DK, Chougaonkar MP (2007) The pre-dose phenomenon in the OSL signal of quartz. Radiat Meas 42(8):1265–1272

Lamothe M (2004) Optical dating of pottery, burnt stones, and sediments from selected Quebec archaeological sites. Can J Earth Sci 41:659–667

Lipo C, Feathers JK, Dunnell RC (2005) Temporal dating requirements, luminescence dates and the resolution of chronological structure of late prehistoric deposits in the central Mississippi River valley. American Antiquity 70:527–544

Liritzis Y, Mckerrell H (1979) Some TL Dates and Neutron Activation Analysis of Greek Neolithic Pottery. Archaeophysika 10:489–501

Liritzis Y (1980) 100°C quartz peak: a new normalization factor. Ancient TL 11:6–7

Liritzis Y (1982) Non-linear TL response of quartz grains; some annealling experiments. P.A.C.T. I(6):209–213

Liritzis Y (1987) The Chernobyl fallout in Greece and its effects on the dating of archaeological materials 260(2–3):534–537

Liritzis I (1994) A new dating method by thermoluminescence of carved megalithic stone building. Comptes Rendus de l 'Academie des Sciences, Paris 319(2):603–610

Liritzis I, Galloway RB, Theocaris PS (1994) Thermoluminescence dating of ceramics revisited: optical stimulated luminescence of quartz single aliquot with green light emitting diodes. J Radioanal; Nucl Chem Lett 188:189–198

Liritzis I (1997) Cluster analysis of thermoluminescence parameters of ancient Hellenic limestone monuments. Naturwissenschaften 84:463–465

Liritzis I, Galloway RB, Hong DG (1997) Single aliquot dating of ceramics by green light stimulation of luminescence from quartz. Nuclear Instruments and Methods B 132:457–467

Liritzis I, Bakopoulos Y (1997) Functional behaviour of solar bleached thermoluminescence in calcities. Nuclear Instruments and Methods B 132:87–92

Liritzis I, Guilbert P, Foti F, Schvoerer M (1997) The Temple of Apollo (Delphi) strengthens new thermoluminescence dating method. Geoarchaeology Int 12(5):479–496

Liritzis I, Galloway RB (1999) Dating implications from solar bleaching of thermoluminescente of ancient marble. J Radioanal Nucl Chem 241(2):361–368

Liritzis I (2000) Advances in thermo- and opto-luminescence dating of environmental materials (sedimentary deposits): part I: techniques. Global Nest: Int J 2(1): 3–27, and Part II: applications. Global Nest 2(1): 29–49)

Liritzis I (2001) Searching for precision of a new 'luminescence clock' in dating calcitic rocks. J. Radioanalytical & Nuclear Chemistry 247(3):727–730

Liritzis I, Vafiadou A (2005) Dating by luminescence of ancient megalithic masonry. Mediterranean Archaeology & Archaeometry 5(1):25–38

Liritzis I, Sideris C, Vafiadou A, Mitsis J (2007) Mineralogical petrological and radioactivity aspects of some building material from Egyptian Old Kingdom monuments. J Cultural Heritage 9:1–13

Liritzis I, Kitis G, Galloway RB, Vafiadou A, Tsirliganis N, Polymeris G (2008) Probing luminescence dating of archaeologically significant carved rock types. Mediterranean Archaeology & Archaeometry 8(1):61–79

Liritzis I (2010) Strofilas (Andros Island, Greece): New evidence of Cycladic Final Neolithic dated by novel luminescence and Obsidian Hydration methods. J Archaeological Science 37:1367–1377

Liritzis I, Polymeris G, Zacharias N (2010a) Surface luminescence dating of 'Dragon Houses' and Armena Gate at Styra (Euboea, Greece). Mediterranean archaeology & archaeometry, Special Issue, (D.Keller, guest editor), 10(3):65–81

Liritzis I, Drivaliari A, Polymeris G, Katagas CH (2010b) New quartz technique for OSL dating of limestones. Mediterranean Archaeology & Archaeometry 10(1)81–87

Liritzis I, Zacharias N, Polymeris GS, Kitis G, Ernstson K, Sudhaus D, Neumair A, Mayer W, Rappenglück MA, Rappenglück B (2010c) The Chiemgau meteorite impact and tsunami event (southest Germany): First OSL dating. Mediterr Archaeol Archaeometry 10(4):17–33

Liritzis I, Laskaris N (2011) A new mathematical approximation of sunlight attenuation in rocks for surface luminescence dating. J Luminescence 131:1874–1884

Mercier M, Valladas H (2003) Reassessment of TL age estimates of burnt flints from the Paleolithic site of Tabun Cavem, Israel. J Hum Evol 45(5):401–409

Mercier N, Wengler L, Valladas H, Joron J-L, Froget L, Reyss JL (2007a) The Rhafas Cave (Morocco): Chronology of the mousterian and aterian archaeological occupations and their implications for Quaternary geochronology based on luminescence (TL/OSL) age determinations. Quat Geochronol 2:309–313

Mercier N, Valladas H, Froget L, Joron J-L, Reyss J-L, Weiner S, Goldberg P, Meignen L, Bar-Yosel O, Belfer-Cohen A, Check M, Kuhn SL, Stiner MC, Tillier A-M, Arensburg B, Vandermeersch B (2007b) Hayonim Cave: a TL-based chronology for this Levantine Mousterian sequence. J Archaeol Sci 34:1064–1077

Meric N, Kosal M, Atlihan M, Rabia Yuce U (2008) OSL properties of anthropological bone and tooth. Radiation Physics and Chemistry 77:685–689

Morganstein ME, Luo S, Ku TL, Feathers JK (2003) Uranium-series and luminescence dating of volcanic lithic artifacts. Archaeometry 45:503–518

Nambi KSV, Murty MLK (1981), Thermoluminescence dating of burnt clay from an upper Paleolithic occupation level from a carve site in south India. Current Science 50(21):945–946

Nakata Y, Tamaki M, Hashimoto T (2007) Red-thermoluminescence dating using quartz grains extracted from a roof-tile of an old Japanese temple. J Radioanal Nucl Chem 272:433–438

Park JS, Nakamura T, Price DM (2005) Estimation of the age of a pre-industrial iron site in Korea. Archaeometry 47:861–868

Polymeris GS, Sakalis A, Papadopoulou D, Dallas G, Kitis G, Tsirliganis NC (2006) Firing temperature of pottery using TL and OSL techniques. Nucl Instrum Meth Phys Res, Sect A 580:747–750

Rendell HM, Townsend PD, Wood RA, Luff BJ (1994) Thermal treatment and emission spectra of TL from quartz. Radiat Meas 23:441–449

Rhodes EJ (2010) Developments in optically stimulated luminescence age control for geoarchaeologcial sediments and hearths in western New South Wales, Australia. Quaternary Geochronology 5:348–352

Rhodes EJ, Fanning PC, Holdaway SJ, Bolton C (2009) Ancient surfaces? Dating archaeological surfaces in western NSW using OSL. In: Fairbairn, A. & O'Connor, S. (eds) New Directions in Archaeological Science. Terra Australia 28:189–200

Richter D, Tosteven G, Škrdla P (2008) Thermoluminescence dating of the type locality of the Bohunician (Brno-Bohinice, Czech Republic). Journal of Human Evolution 55:871–885

Richter D (2007) Advantages and limitations of thermoluminescence dating of heated flint from Paleolithic sites. Geoarchaeology 22:671–683

Richter D, Krbetschek M (2006) A new thermoluminescence dating technique for heated flint. Archaeometry 48:695–715

Richter D, Temming H (2006) Testing heated flint palaeodose protocols using dose recovery procedures. Radiat Meas 41:819–825

Richter D, Tostevin G, Škrdla P, Davies W (2009) New radiometric ages for the early upper palaeolithic type locality of Brno-Bohunice (Czech Republic): comparison of OSL, IRSL, TL and 14C dating results. J Archaeol Sci 36:708–720

Roberts RG (1997) Luminescence dating in archaeology: from origins to optical. Radiat Meas 27:819–892

Sampson EH, Fleming SJ, Bray W (1972) Thermoluminescent dating of Colombian pottery in the Yotoco style. Archaeometry 14:121–128

Sampson CG, Bailiff I, Barnett S (1997) Thermoluminescence dates from Later Stone Age pottery on surface sites in the Upper Karoo. South African Archaeological Bulletin 52:38–42

Sánchez J, Mosquera D, Fenollós JM (2008) TL and OSL Dating of Sediment and Pottery from Two Syrian Archaeological Sites. Geochronometria 31(1):21–29

Sanderson DCW, Bishop P, Stark MT, Spencer JQ (2003) Luminescence dating of anthropogenically reset canal sediments from Angkor Borei, Mekong Delta, Cambodia, Quaternary Science Reviews 22, 1111–1121

Sanderson DCW, Bishop P, Stark M, Alexander S, Penny D (2007) Luminescence dating of canal sediments from Angkor Borei, Mekong Delta, Southern Cambodia. Quaternary Geochronology 2:322-329

Simms AR, Kouremenos P, DeWitt R, Drewry AM (2011) A New Approach to Reconstructing Sea Levels in Antarctica Using Optically Stimulated Luminescence of Cobble Surfaces. Quaternary Geochronology 6:50–60

Singhvi AK, Sharma YP, Agrawal DP (1982) Thermoluminescence dating of sand dunes in Rajasthan, India. Nature 295:313–315

Singhvi AK, Sengupta D, Deraniyagala SU (1986a) Thermoluminescence Dating of quaternary red sand beds: a case study of coastal dunes in Sri Lanka. Earth and Planetary Science Letters 80:139

Singhvi AK, Sauer W, Wagner GA, Kroger K (1986b) Thermoluminescence Dating of Loess at Plaidter Hummerich and its implications on the chronology of Neanderthal man. Naturwissenschaften 73:205

Singhvi AK, Chawla S (1986c) Thermoluminescence dating of non-pottery materials, implications for the chronology of the Early Man. J Arch Chem 4:59

Singhvi AK, Chawla S (1989) Thermoluminescence dating of Archaeological Sediments S. Chawla and A.K. Singhvi. Naturwissen schaften 76(9):416

Singhvi AK, Agrawal BP, Nambi KSV (1991) Thermoluminescence dating: An update on applications to Indian Archaeology. Recent advances in Marine Arch, (Ed. S.R. Rao), Society for Marine Archaeology, N10, Goa, 173–180

Sohbati R, Murray AS, Jain M, Buylaert JP, Thomsen KJ (2011) Investigating the resetting of OSL signals in rock surfaces. Geochronometria 38:249–258

Sohbati R, Murray A, Buylaert JP, Almeida N, Cunha PP (2012) Optically stimulated luminescence (OSL) dating of quartzite cobbles from the Tapada do Montinho archaeological site (east-central Portugal). Boreas 41:452–462

Sunta CM, David M (1982) Firing temperatures of pottery from predose sensitization of TL. PACT J 6:460–467

Susino GJ (2010) Optical dating and microwaste – archaeological applications. Quaternary Geochronology 5:306–310

Thomas PJ, Nagabhushanam P, Reddy DV (2008) Optically stimulated luminescence dating of heated materials using single-aliquot regenerative-dose procedure: a feasibility study using archaeological artefacts from India. J Archaeol Sci 35:781–790

Theocaris P, Liritzis I, Lagios E, Sampson A (1997) Geophysical prospection and archaeological test excavation and dating in two Hellenic pyramids. Surveys in Geophysics 17:593–618

Tribolo C, Mercier N, Selo M, Valladas H, Joron JL, Reyss JL, Henshilwood C, Yates R (2006) TL dating of burnt lithics from Blombos Cave (South Africa): Further evidence for the antiquity of modern human behavior. Archaeometry 48:341–357

Tribolo C, Mercier N, Valladas H, Joron J-L, Guibert P, Lefrais Y, Selo M, Texier P-J, Rigaud J-Ph, Porraz G, Poggenpoel C, Parkington J, Texier J-P, Lenoble A (2009)

Thermoluminescence dating of a Stillbay–Howiesons Poort sequence at Diepkloof Rock Shelter (Western Cape, South Africa). J Archaeol Sci 36:730–739

Vafiadou A, Murray AS, Liritzis I (2007) Optically Stimulated Luminescence (OSL) dating investigations of rock and underlying soil from three case studies. J Archaeol Sci 34:1659–1669

Valladas H, Wadley L, Mercier N, Froget L, Tribolo C, Reyss J-L, Joron J-L (2005) Thermoluminescence dating on burnt lithics from Middle Stone Age layers at Rose Cottage Cave. S Afr J Sci 101:169–174

Valladas H, Mercier N, Froget L, Joron J-L, Reyss J-L, Karkanas P, Panagopoulou E, Kyparissi-Apostolika N (2007) TL age-estimates for the Middle Palaeolithic layers at Theopetra cave (Greece). Quat Geochronol 2:303–308

Valladas H, Mercier N, Ayliffe LK, Falguères C, Bahain J-J, Dolo JM, Froget L, Joron J-L, Masaoudi H, Reyss J-L, Moncel M-H (2008) Radiometric dates for the Middle Palaeolithic sequence of Payre (Ardèche, France). Quat Geochronol 3:377–389

Vieillevigne E, Guibert P, Zuccarello AR, Bechtel F (2006) The potential of optically stimulated luminescence for medieval building: a case study at Termez, Uzbekistan. Rad Meas 41:991–994

Vieillevigne E, Guibert P, Bechtel F (2007) Luminescence chronology of the medieval citadel of Termez, Uzbekistan : TL dating of bricks masonries. J Archaeological Science 34:1402–1416

Vaz JE (1983) The effect of insolation on the thermoluminescence response of an archeological stone sculpture. PACT J 9:335–342

Wang W-D (2008) Evaluation for the paleodose in thermoluminescence dating of porcelain. Sci China Ser E: Technol Sci 51:260–267

Wang XL, Lu YC, Wintle AG (2006) Recuperated OSL dating of fine-grained quartz in Chinese loess. Quat Geochronol 1:89–100

Whittle EH, Arnaud JM (1975) Theermoluminescence dating of Neolithic and Chalcolithic pottery from sites in central Portugal. Archaeometry 17(1):5–24

Wintle AG, Huntley DJ (1980) Thermoluminescence dating of ocean sediments. Canadian Journal of Earth Sciences 17:348–360

Yang X-Y, Kadereit A, Wagner GA, Wagner I, Zhang J–Z (2005) TL and IRSL dating of Jiahu relics and sediments: clue of 7th millennium BC civilization in central China. J Archaeol Sci 32:1045–1051

Zacharias N, Mauz B, Michael CT (2002) Luminescence quartz dating of lime mortars. A first research approach, Radiation Protection Dosimetry 101(1–4):379–382

Zacharias N, Michael CT, Philaniotou-Hadjiannastasiou O, Hein A, Bassiakos Y (2006a) Fine-grain TL dating of archaeometallurgicalfurmace walls. Journal of Cultural Heritage 7:23–29

Zacharias N, Michael CT, Georgakopoulou M, Kilikoglou V, Bassiakos Y (2006b) Quartz TL dating on selected layers from archaeometallurgical kiln fragments: A proposed procedure to overcome age dispersion. Geochronometria 25:29–36

Zacharias N, Schwedt A, Buxeda I, Garrigøs J, Michael CT, Mommsen H, Kilikoglou V (2007) A contribution to the study of post-depositional alterations of pottery using TL dating analysis. J Archaeol Sci 34:1804–1809

Zacharias N, Beltsios K, Oikonomou Ar, Karydas AG, Aravantinos V, Bassiakos Y (2008) Thermally and optically stimulated luminescence of an archaeological glass collection from Thebes, Greece. J Non-crystalline Solids 354:761–767

Zimmerman J (1971) The radiation-induced increase of the 110°C thermoluminescence sensitivity of fired quartz. J Phys C: Solid State Phys 4:3265–3276

Chapter 5
Luminescence-Based Authenticity Testing

Luminescence can also be employed as a method for authenticating ceramic art materials. Counterfeiting pottery is a multi-million dollars business, and the fakes have been around since time immemorial. Authenticity analysis begins with a stylistical and technological study of the artifact, but can be supplemented by luminescence work on a small portion ($\sim$50–10 mg total) drawn from an inconspicuous part of the specimen by drilling with a tungestan carbide bur in subdued light. Such amounts on the order of milligrams is adequate in most authenticity applications, not only for equivalent dose determinations but also for dose rate measurements. In such studies De is preferably measured by single aliquot protocols tailored for TL (Michael and Zacharias 2006) Generally, environmental dose-rates can only be guessed at because of lack of information on where all the pieces of an artefact have been. This however, does not affect the conclusions as it constitutes a minor fraction of the total dose. The main TL techniques of authenticity of fired clay fabric were developed during 1970s and include the pre-dose, the subtraction (between fine-grain and inclusion), and the zircon (Fleming and Stoneham 1971; Fleming 1971; Sutton and Zimmerman 1976).

Regarding the methods of De determination, the measured principle and method for the paleodose of porcelains has been approached by the exponential regression method, which is found more suitable for porcelain dating, and it has advantages compared with the standard pre-dose method (i.e. linear regression method) (Wang et al. 2006). Thirty-nine shards and porcelains from past dynasties of China were examined Their experimental results have shown show that the measured errors are 15 % ($\pm 1\sigma$) for the paleodose and 17 % ($\pm 1\sigma$) for the annual dose respectively, while the TL age error is about 23 % ($\pm 1\sigma$) in this method. The larger Chinese porcelains from the museums and the nation-wide collectors have been dated by this method. The results show that the certainty about the authenticity testing is larger than 95 %, and the measurable porcelains make up about 95 % of the porcelain dated. It was very successful in discrimination for the imitations of ancient Chinese porcelains.

Some early examples include Chinese ceramics and some important young terracottas, including forgeries of Etruscan material. There, the thermoluminescence of

I. Liritzis et al., *Luminescence Dating in Archaeology, Anthropology, and Geoarchaeology*, SpringerBriefs in Earth System Sciences,
DOI: 10.1007/978-3-319-00170-8_5,

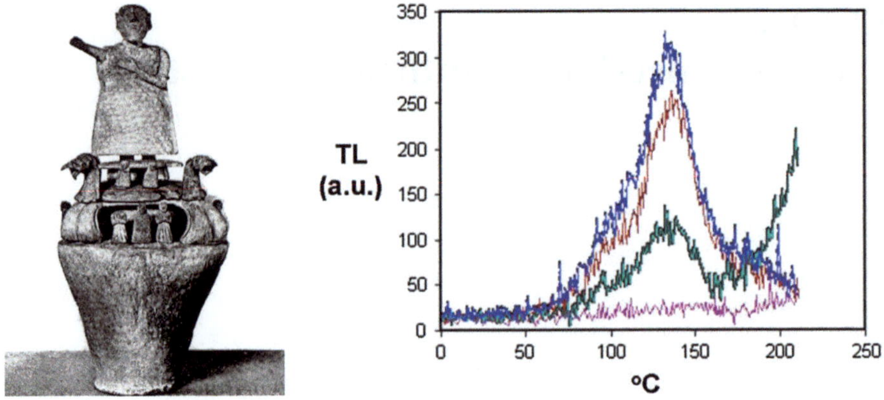

Fig. 5.1 Photo of the Forthigham Vase (Washington University Museum); the 110 °C TL sensitivity glows are shown, monitored during the application of the pre-dose protocol by Bailiff (1994) ranging from the initial (lower curve) to the final response (higher one)

the quartz peak at around 110 °C was used which changes sensitivity dramatically under the combined influence of radiation and heat. Laboratory heating of ancient pottery induces TL sensitivity changes in the quartz of that pottery which is a measure of the natural radiation dose experienced during burial. Quantitative assessment of that archaeological dose (through comparison with the sensitivity change caused by a known laboratory dose) together with measurement of the radioactive content of the pottery lead to their age determination (Fleming 1972).

A study on the authenticity of the Tanagrae figurines of the Louvre Museum Zink and Porto (2005) was based on measurements of the environmental dose parameters of Boeotia (Central Greece), the place of origin of the figurines. The study showed that 10 % of the 140 examined objects were fakes. Figure 5.1 shows the TL glow recorded during authenticity testing for the Washington University Museum of a vase and figures for which a typological age of 2,600 years was expected (Etruscan Period). Nine items were tested, and 7 yielded a zero age, while only two snake-like forms gave signals that conformed to the expected age. The TL examination showed that the vase is a modern production mixing new materials with those of archaeological value (Fig. 5.1).

In a modern study (Mavrokefalos 2007) a clay unfired briquette was broken in several parts; each part was subjected to a single heating and TL recording cycle. The temperatures applied ranged from 80 °C and up to 300 °C. The study is evident that heating clay at temperatures lower than 300 °C and for 1 hour of duration do not zero the high temperature TL peaks. Thus, for modern replicas of archaeological objects, the combined application of moderate heating temperatures for short periods of time could produce TL signals similar to the 'archaeological' ones. To avoid such implications, an estimation of the temperature the artefact was exposed, using a fast, established and quasi-destructive technique like e.g. Scanning Electron Microscopy, FTIR or RAMAN, should be included as a standard step in routine authenticity testing.

Theoretically, it is difficult to "fool" a TL test as it is difficult to replicate natural radiation field comprising grains dosed to variable degrees. Normally the fine grains will have higher dose compared to coarse grains and hence the dose determination of fine and coarse fraction provides a straightforward means to detect a forgery as it is difficult to obtain this difference in a fake sample. Similarly the depth dependence of dose is another indicator that can be used to differentiate between a fake and a real object. A real object will not have this dependence but an irradiated fake will show a finite depth dependence and this can be determined easily.

References

Bailiff IK (1994) The pre-dose technique. Radiat Meas 23(2–3):471–479

Fleming SJ (1971) New techniques of thermoluminescent dating of ancient pottery: the predose method. In: Mejdahl V (ed) Proceedings of the third international conference on luminescence dosimetry, held at the Danish AEC research establishment Risø, Riso report No. 249, 895–929, 11–14 Oct 1971

Fleming SJ, Stoneham D (1971) New techniques of thermoluminescent dating of ancient pottery: the substraction method. In: Mejdahl V (ed) Proceedings of the third international conference on luminescence dosimetry, held at the Danish AEC research establishment Risø, Riso report No. 249, 880–897, 11–14 Oct 1971

Fleming SJ (1972) Thermoluminescence authenticity testing of ancient ceramics using radiation-sensitivity changes in quartz. Naturwissenschaften 1972:145–151

Mavrokefalos N (2007) Study of the TL signals versus annealing temperatures of ceramic objects, Diploma Thesis (in Greek), Department of Chemical-Engineering, N.T.U.A

Michael CT, Zacharias N (2006) Equivalent dose estimation in TL dating using a single-aliquot of polymineral fine grains. Radiat Prot Dosim 119(1–4):458–461

Sutton SR, Zimmerman DW (1976) Thermoluminescent dating using zircon grains from archaeological ceramics. Archaeometry 18:125–134

Wang W, Xia J, Zhou Z (2006) Thermoluminescence dating of the ancient chinese porcelain using a regression method of saturation exponential in pre-dose technique. Sci China: Series E Technol Sci 49(2):194–209

Zink A, Porto E (2005) Luminescence dating of the Tanagra terracottas of the louvre collections. Geochronometria 24:21–26

Chapter 6
Luminescence Dating in Geomorphological and Geoarchaeological Research in Europe: Application Examples

6.1 Sediments: The Rationale and the Valuable Information

As for any historically oriented science, well established chronometries are a fundamental prerequisite in geomorphology and geoarchaeology in order to determine the temporal coincidences of events and to reveal possible causes and effects. Geomorphology deals with the natural processes, forms and substrates shaping the earth's surface and landscapes in different climates. Thus, e.g., dunes are built up by aeolian processes in arid landscapes, where wind mobilizes sand on unvegetated ground and redeposits at suitable location, where the wind vector drops. In humid climates river networks develop, draining the landscapes by fluvial processes which leave behind distinctive geomorphic forms, as, e.g., point bars and alluvial plains made up of gravelly, sandy, silty and/or clayey material. From natural archives like colluvial sediments, sand-dunes or alluvial plains geomorphologists reconstruct former landscapes and the processes that shaped them. Natural changes of climate and landscapes over time act as constraining factors and driving forces affecting also the rise and fall of cultures. Thus, e.g., the increasing aridity of northern Africa, where people had lived widely scattered over extended areas in more humid times of the Holocene warm period, seems to have been a driving force for the development of the ancient civilisation of Egypt along the lower reaches of the river Nile (Kuper and Kröpelin 2006), an allochthonous river which became the most important water- and lifeline for the people ~5 ka ago.

While natural landscapes and climate changes have an impact on people, humans by removing the natural vegetation cover and by cultivating the land may have an equally strong impact on the appearance of a landscape. Thus, an increasing land-use demand may lead to uncovered surfaces invoking aeolian processes and the accumulation of sand-dunes, which, unlike those owned to natural climate-induced aridisation, are due to man-made desertification (Kar et al. 1998).

Interactions of humans and landscapes of the past is the research field of geoarchaeology. Using geoscientific methods like geomorphology, geoarchaeology uncovers the former human-landscape interaction.

I. Liritzis et al., *Luminescence Dating in Archaeology, Anthropology, and Geoarchaeology*, SpringerBriefs in Earth System Sciences,
DOI: 10.1007/978-3-319-00170-8_6,

Chronometries for landforms could be established using diagnostic artifacts or radiocarbon. Necessary preconditions are that respective archaeological and organic material is available in the first place and that it is preserved *in situ* and had not been reworked after the event that is dated. However, these prerequisites are not always fulfilled. Therefore it might be advantageous to date the sediments directly by luminescence. A key element in the dating of sediments is the energy and duration of day light exposure seen by the mineral grains prior to their deposition as sediment. Thus for example subareal transport under clear day light implies rapid bleaching (seconds to minutes) and the same under dusty condition would imply truncated day light flux and hence slower, and possibly partial bleaching. In the case of subaquatic transport, the depth of the water column and the sediment load determine the net energy and the flux of light seen by the individual mineral grains. Grains transported in suspension under turbulent condition may be better bleached than the bed load beneath a muddy water column. Important here is to realize that the efficiency of bleaching depends on the wavelength (Berger 1990). The bleaching rate depends also on the mineral type. Thus e.g. quartz bleaches faster than the feldspars on blue light stimulation and feldspar bleaches with IR (Hutt et al. 1988) and quartz does not. However, at higher temperature (> 120 °C) quartz is bleached by IR (Polymeris et al. 2008).

If not properly analyzed, incompletely bleached sediments, would deliver maximum ages (*terminus post quem*). Tests of incomplete (partial, differential, e.g. Duller 1991, 1994) bleaching can be carried out by checking the trend of the luminescence signal with continuing stimulation time (D_e-plateau test e.g. Rees-Jones and Tite 1997) or, in the case of coarse grains, by scrutinizing the distribution of D_e-values of a greater number of small aliquots (Li 1994) or the dose distribution of single grains (Olley et al. 1999; Thomsen et al. 2007) or by using special analysis techniques (e.g. Singhvi and Lang 1998) (Fig. 6.1).

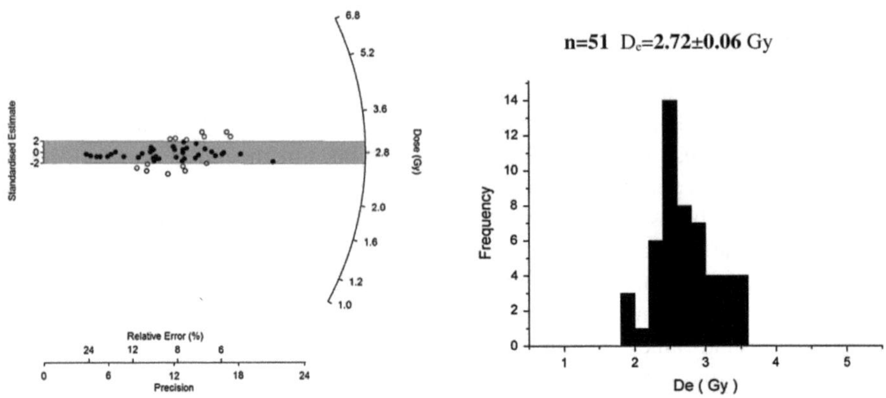

Fig. 6.1 Radial-plot and D_e distribution of quartz aliquots from an 8 ka old loess sample. The results suggest that the sample was well bleached prior to deposition as it shows a normal D_e distribution and their D_e values concentrate well in the radial-plot. In radial-plots, D_e within the shaded region (<2 σ) are represented by filled points (Li 1994)

A version of TL, the beta-TL dating, for sediments is used alternatively, based on an equation relating D_e and K content only, especially when U, Th radioactivity data are not readily available (Liritzis 1989).

As no efficient way exists to extract pure feldspar fine-grains, feldspar fine-grain dating is done on polymineral fine-grains. Unlike quartz, feldspars may be stimulated by light in the near infrared (IRSL), which allows discrimination of the feldspar component of a polymineral separate. In contrast to quartz, however, which usually shows a stable luminescence signal, feldspar may exhibit *anomalous fading* of the luminescence signal (Wintle 1973). Therefore, optical dating of feldspar has to include tests scrutinizing the signal stability of every dated sample and a fading correction test, such as those by Huntley and Lamothe (2001).

Colluvial and Fluvial Deposits

In the vicinity of archaeological sites, colluvial and alluvial deposits are important sediment archives of information on the landscape-disturbing activities of man in historical and prehistorical times. Colluvial deposits are found on footslopes and in nearby depressions as, e.g., in adjacent dry valleys. These deposits represent an accumulation from soil erosion, which occurred in upslope direction, the sediment source area usually being only few tens to several hundred metres away from the sediment sink so that connections to the source and the archaeological site may be inferred relatively clearly from a colluvial deposit—if both the site and the colluvium can be reliably dated. Soil erosion and colluvium formation occur if man clears or destroys the natural vegetation cover so that processes of sediment mobilization (induced by wind, rainfall and snowmelt and enhanced by plough erosion), attains a level above the natural background. A part of the mobilized soil material on the slopes is trapped in close-by sediment sinks and the rest is transferred to a coupled creek or river, which transfers part of it to the next discharging river and so forth. Such remobilized sediments record the past human interference in the landscape. A reliable chronometry is then needed to elucidate the history of man-landscape interactions. ^{14}C-dating and pottery typology are often not the only choices, OSL dating of colluvial sediments contributes significantly to the chronological problem, as is illustrated by Fig. 6.2 for profile K1. It concerns an example from the footslope colluvium at the Neolithic site Vaihingen, upon the river Enz, S.Germany. Here, colluvial sediments accumulated on the lower slopes of a ridge as a result of past and present soil erosion. This allowed reconstruction of the depositional history of two colluvial profiles (K1 and K2) and also to identify temporary sedimentary sinks in its transportational pathway. For the underlying solifluction layer in colluvial profile K2 an OSL-age of 10.8 $\pm$ 1.1 ka was obtained.

With respect to luminescence dating, bleaching conditions are often not optimal for colluvial and alluvial deposits, as transport distances may be very short as reported e.g. by Kadereit et al. (2006a), and turbid water flow may cause sediment transport. Colluvial infillings of narrow archaeological structures like ditches may contain sediment lumps tumbled off the walls of an earthwork, the interior

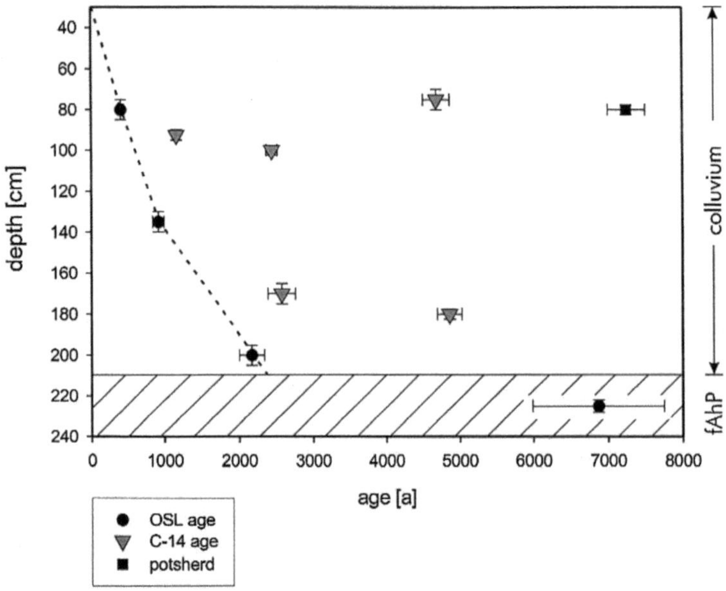

Fig. 6.2 Numeric ages versus depth below ground level of a footslope colluvium at the Neolithic site of Vaihingen upon the river Enz in S. Germany (from Lang and Hönscheidt 1999)

sediment grains of which have not received any daylight during the relocation (e.g. Lang et al. 1999). Also, calcareous soil sediments or otherwise agglutinating grains may be transported as peds which cannot be bleached inside and therefore deliver only maximum ages for the corresponding periods of past soil erosion (Kadereit et al. 2010). Apart from such extreme conditions, recent progress in the luminescence techniques generally allows the successful optical dating of colluvial (e.g. Fuchs and Lang 2009) and fluvial deposits (e.g. Wallinga 2008).

Several studies on colluvial deposits in the agriculturally favourable loess covered areas of southern Germany, which had been occupied in all known cultural periods since the onset of agriculture in the Linear Band Ceramic culture period ca. 7.5 ka ago, show that man-induced colluviation occurred as early as the Neolithic (e.g. Lang and Wagner 1996), increased in Bronze Age times (e.g. Lang et al. 2001), and has exhibited a significant relief-shaping impact on the landscape since Iron Age times and Medieval times (e.g. Kadereit et al. 2010). Remarkably, at some locations Early Bronze Age colluvium is the only evidence of human presence ~3.6 ka ago, probably because building foundations were too shallow for archaeological remains to being preserved sensu stricto. Most dating of the loess-borne colluvium was done by IRSL of the polymineral fine-grain fraction of the material, often using a MAAD protocol.

Recently, Fuchs et al. (2011) investigated the catchment (~100 km^2) of the river Aufsess in the loess-free and therefore only later (late-Neolithic times ca. 5.1 ka ago) settled area of SE-Germany. Intensive dating of both colluvial and alluvial sediments using a BLSL SAR-protocol on quartz coarse-grains suggests non-isochronical behavior of the two sediment sink types, with the colluvial system being deposited in late-Neolithic and in Modern times ~1500 AD. In the same period the alluvial system shows declining sedimentation rates, with the alluvial system first responding 2–3 ka after the late Neolithic colluviation period and showing a response peak in the Middle Ages ~1000 AD, when colluviation is relatively low. The results also show a ratio of ~60:10:30 of sediments stored as colluvium to those stored as alluvium to those transported out of the catchment. Although obviously superimposed by a temporally varying connectivity between the colluvial and the discharging alluvial system, man-landscape interactions are identified as the triggering factors of the changes in the (pre-)historic sediment budgets (Fuchs et al. (2011).

For a study site in central Belgium Rommens et al. (2007) document first colluvium formation in Iron Age times. Sedimentation rates of 2.9 ± 0.9 tons per hectare per year double to 5.2 ± 1.5 t/ha/a in the Roman Period and triple again to 18.0 ± 2 t/ha/a in the Middle Ages, a result analogous to that in Germany. The authors argue that low-magnitude high-frequency soil-erosion and colluviation events which allow prolonged periods of soil cultivation for additional bleaching of the sediment grains after deposition, favour the successful dating of colluvial deposits with luminescence techniques.

Whereas in central and western Europe colluviation has effectively shaped the topography of a landscape mainly since Iron Age times—partly because of the later reworking of older colluvium (Lang and Hönscheidt 1999), Fuchs et al. (2004) show for a study site in the basin of Phlious on the Peloponnesos/Greece that soil erosion induced colluvium formation in the Mediterranean area experienced increased sedimentation rates in the Neolithic period, i.e. at the time whem people started to settle the area. The distinct response recorded in the sediment sinks might be explained by a contemporaneous increase in precipitation (Fuchs and Wagner 2005; Fuchs 2007). Therefore, the study shows impressively how man may provide the general set-up for the natural geomorphic processes to shape a landscape lastingly. The study identifies the Middle and Late Bronze Age times, the Roman period and the Modern period as times of exceptionally high colluvium formation. The high resolution chronometry is based on the BLSL dating of small aliquots (200–500 grains) of quartz coarse grains using the SAR protocol.

At the end of the last glacial period, i.e. during the Weichselian Upper-Pleniglacial and especially the Younger Dryas cold reversal, aeolian mobilization of sand south of the retreating Scandinavian ice sheet left the European Sand Belt (ESB) stretching from Britain to the Polish-Russian border and made up of sand dunes as well as sheet like sandy cover beds. Using both large (1000–2000 grains) and small aliquots (<100 grains) and a robust BLSL SAR-protocol for coarse-grain quartz that had been tested against ^{14}C-dated palaeosoils of the ESB (e.g. Kaiser et al. 2009), Hilgers (2007) confirmed the remobilization of sands in Holocene times. Periods of dune reactivation are around 6.3, 4.5, 3.5, 2.6 ka ago and 300, 1100, 1300 and 1700 AD. Unlike the late Pleistocene dune formation, Holocene dune activity is not isochronal at all investigated sites in northern

Germany. This non-synchronicity plus the fact that there is no climatic impact that could explain vegetation deterioration causing the aeolian sediment mobilization in Holocene times leaves human activity as the likely cause of wind-borne soil erosion during the different settlement periods (Hilgers 2007).

A yet unresolved question for settlement mounds (tells) in the northern temperate Balkans is why they appeared later in time and disappeared earlier than their counterparts in the southern Mediterranean Balkans (Schier and Drasovean 2004). For a representative Late Neolthic/Early Copper Age tell site at Uivar in Romania, optical dating (IRSL MAA dating of polymineral fine grains and BLSL SAR-dating of quartz coarse grain) of fluvial deposits burying and leveling the former fluvial landscape around the prehistoric settlement site shows that the river Bega was still aggrading its river bed until ~6.5 ka ago, likely due to a climatic impulse, and that preservation of settlement remains and tell emergence could occur only after that time (Kadereit et al. 2006b). These results were possible even though the sediments had been apparently insufficiently bleached prior to deposition.

The Palpa-Nasca area, S.Peru, had been considered as geomorphologically stable ever since the famous giant geoglyphs had been created in the presently hyper-arid northern Atacama desert landscape by the people of the Nasca (~200 BC–600 AD) and pre-Nasca cultures (Eitel et al. 2005). However, BLSL SAR-dating of coarse-grain quartz from fluvial deposits of local zero-order creeks and allochthonous rivers draining the western Andean flank, together with ^{14}C-dating, helped to identify the Little Ice Age period (fourteenth–seventeenth century) as a time of major sediment reworking (Unkel et al. 2007).

6.2 Some (Geo-) Archaeological Applications

Stratigraphical integrity means that uncovered artefacts were *deposited* at around the same time as the associated sediment and this aspect should be assessed in every case. An example is a Neolithic floor overlain by a round cobble (millstone) found in situ in an undisturbed state. Both were last exposed to sunlight during destruction of the house at Ftelia, Mykonos and dated to c. 5000 B.C. (Vafiadou et al. 2007). Single-grain dating to assess stratigraphic integrity were employed at Middle Stone Age sites in Africa (e.g., Jacobs et al. 2006; Jacobs 2008a, b), paleoindian sites in the Americas (e.g., Araujo et al. 2008; Feathers et al. 2006a, b), ~20–40 ka sites in Australia (e.g., Bowler et al. 2003; Cupper and Duncan 2006; Olley et al. 2006; David et al. 2007), and in Borneo (Stevens et al. 2007) and at a microlithic site in Mali where it was unclear whether the sediments were Holocene or Pleistocene (Tribolo et al. 2010). Where sediments are mixed, finite mixture models have been employed to distinguish individual-age components; sometimes the component accounting for most grains being taken as the best age for the deposit (e.g., Jacobs et al. 2008). In other cases comparison with radiocarbon dates, suggests that this premise might not always hold (Rhodes et al. 2009). Using single-grain dating at several sites in South Africa following the same protocol, Jacobs et al. (2008) were able to tightly constrain the dates of the Howiesons Poort industry and to a lesser extent, the Still Bay industry. They showed the two industries had relatively brief durations and were separated in time. Neither seemed to correlate with environmental variables. Luminescence dating can also be used to understand cultural sequences and stratigraphy, where these are not clear or are open to different interpretations (e.g., Anderson et al. 2006).

Sediment dating by luminescence has also proven useful for providing a chronology of changes in the regional landscape that correlate with human settlement. Fuchs et al. (2008) dated dunes and underlying calcrete near Cape Town to show how human use of the landscape changed through time. Work in Australia (Rhodes et al. 2009; Fanning et al. 2008) has used OSL on sediments and radiocarbon on hearths to date land forms to see how long a particular surface has been available for accumulation of artifacts. Results showed that the visible surface record, including the quantity and range of artifacts, is largely a function of the age of the surfaces, and that not all artifacts currently on the surface are the same age. These results challenged a conventional view that population and social complexity increased through time, because older records have probably been eroded or buried. A similar evolution was suggested for colluvial deposits, with older deposits being reworked into younger colluvium during subsequent periods of anthropogenically induced geomorphic activity (Lang and Hönscheidt 2003). As luminescnce dating can only determine the last time of sediment reworking, this interpretation was deduced from discrepancies of older, and therefore apparently remobilized charcoal, in younger sediment deposit (cf. figure 6.3). The charcoal, which was taken as an indication of human burn-activity, originated from different periods.

Other examples include Wilkinson et al. (2005), Thomsen et al. (2007), Holliday et al. (2006), Singhvi et al. (2010) where the ages of paleolithic and other tools were assigned based on the dating of the burial strata. OSL or IRSL has also been used to correlate human settlement and activity with changes in sedimentation, e.g., the chronology of aggradation and erosion, and associated human settlement, in the Belan Valley, India (Gibling et al. 2008; William et al. 2006), the dating of drift sands that probably caused abandonment of a medieval settlement in Belgium (Derese et al. 2009), correlation of increased sedimentation with human deforestation in Netherlands (de Moor et al. 2008), correlation of lacustrine sediments with human settlements in Malawi (Scholz et al. 2007), correlation of lacustrine and fluvial sediments with human settlements in Egypt (Bubenzer et al. 2007), correlation of settlement and mobility with changes in river aggradation in Kenya (Wright et al. 2007).

Extending the age range back for sediments is currently being researched intensively. Quartz is generally limited to sediments less than 100 ka or so because of saturation, but research into the signal from TT-OSL (cf. D_e protocols) has shown higher saturation. Therefore much older ages can be obtained. Some archaeological examples include 100–450 ka ages on quartz for a Mousterian site in France (Sun et al. 2010) and 200 ka ages on quartz at a Korean Acheullian site (Kim et al. 2010). Low dose rates also allow older ages on quartz. Reliable dates of 300 ka, but possible ages up to 900 ka, were obtained on sediments associated with Paleolithic industries near Casablanca (Rhodes et al. 2006). A surprisingly early date on quartz of 450 ka, recovered with not a particularly low dose rate since the equivalent dose estimated was 300 Gy, was obtained from Middle Stone Age (MSA) deposits at Kathu Pan in South Africa (Porat et al. 2010). Unavoidably we here turn again to methodological issues, concerning the older ages and inherent thorns of fading. Feldspar has higher saturation limits and therefore has a much older age range, but it is useful for old samples only if anomalous fading, an athermal loss of signal that affects feldspars, can be accounted for. While correcting for anomalous fading appears straightforward for younger samples where the growth curve is in the linear region, no procedure has been agreed upon for older samples, where the fading rate's dependence on dose is significant. Circumventing fading by isolating a signal less prone to fading or developing a reliable correction procedure for older samples is an area of active research. An archaeological example, using IRSL, is the dating of 60 ka Howiesons Poort deposits in South Africa using the dose correction method of Lamothe et al. (2003) (Barre and Lamothe 2010). Others have used IRSL on fine-grained material, using the 420 nm emission band of K-feldspar, on the assumption that this signal is relatively stable in fine grains (e.g., Berger et al. 2008; Wright et al. 2007). Berger et al. (2008) obtained dates up to 250 ka using IRSL on fine grains at Atapuerca, a hominin site in Spain, but got dates up to 950 ka using TL on the same size fraction. Ongoing research is focussed on using high temperature stimulation of feldspars to isolate a non- or low-fading component (Buylaert et al. 2009; Jain and Anjaergaard 2010; Li and Li 2011), while Li and Li (2012) dated loess from China up to 300 ka using IRSL from K-feldspar with high temperature stimulation.

At the other end of the scale, luminescence has proven useful for very young samples, particularly the last 500 years where radiocarbon dating is imprecise due to multiple intercepts in the calibration curve (e.g., Derese et al. 2009; Borejsza et al. 2008; Feathers 2008).

A disadvantage of quartz for luminescence dating is that in many regions of the world the quartz UV emission shows very low luminescence sensitivity. This necessitates use of feldspars or a different emission band for quartz. The red emission band from TL of quartz has been employed in islands southeast of Asia, where quartz sensitivity is also low. Westaway et al. (2007) developed a dual-aliquot regenerative dose (DAP) procedure to date sediments at the well-known hominid site on Flores Island in the 100–200 ka range. The procedure was also used to date a 70 ka archaeological site on Java (Morwood et al. 2008).

Another recent, but rather unusual application should be mentioned: the dating of sediments overlying and enclosing human footprints. Dates have been reported for footprints of 120 ka in Africa (Jacobs et al. 2009), of 20 ka in Australia (Webb et al. 2006), and of 36 ka in Mexico (Gonzales et al. 2006). The last is rather controversial and the results have been disputed (Duller 2006).

Burbidge et al. (2007) recommended "luminescence profiling" for complex situations, as, e.g., mixing, poor bleaching, or unsuitable signals. In a first campaign, a number of test samples are collected and these are analyzed to find out which locations are likely to provide the most reliable dates. In a following campaign, dating samples are collected from these locations. This procedure can be streamlined for one field season by use of portable luminescence readers (Sanderson and Murphy 2010). In one application in trying to date sediments from a Neolithic ditch in Italy, Sanderson and Murphy (2009) discovered in the field, that the upper fill contained a large residual signal, the sediments probably stemming from original unbleached backfill, while the lower fill provided signals suitable for dating. This was later confirmed by laboratory measurements.

6.3 Accuracy and Reliability in Sediment Dating

When dating archaeological sites or objects, it is often advantageous to use several dating methods in parallel which deliver results that are independent from each other. Also, researchers improving luminescence dating protocols and developing new techniques aim to test the quality of a dating method against data that are available from preferably independent dating techniques. One such study (Hatté et al.2001; Lang et al. 2003) was performed at the site of Nussloch in S-Germany. This site is famous for its thick late-Pleistocene loess deposits that contain several palaeosoils of varying intensity and exhibit a variety of other proxies that allow detailed palaeoclimatic and palaeoenvironmental reconstruction. The site is considered a type sequence for Western Europe (Antoine et al. 2009; Rousseau et al. 2002). For southern Germany a terrestrial palaeoclimatic archive of such high resolution for the period of the last glacial maximum (LGM) is especially valuable with regard to the Palaeolithic art objects (e.g. bone and ivory flutes and ivory venus of Hohle Fels cave) found in karst caves of the nearby Swabian Alb in layers from the earliest Aurignacien (base of Upper Palaeolithic) onwards which give evidence of a sudden cultural blooming at the time of the appearance of anatomically modern humans in the area >35–40 ka ago (e.g. Conard et al. 2009; Rau et al. 2009). At the Nussloch section undiagnostic Middle or Upper Palaeolithic remains of stone artefacts and animal bones were found in channel fills preserved within a thermokarst depression (Kind 2000). For Nussloch, a robust chronometry was established using both IRSL-dating of polymineral fine-grains and accelerator mass spectrometry (AMS) ^{14}C-dating of in situ organic macro-remnants and humin fractions extracted from the sediments (Lang et al. 2003). IRSL-dating was performed applying a MAAD protocol and using the luminescence signal of the blue (410 nm) emission of the feldspar component of the polymineral fine-grains extracted from the loess deposits.

The upper 12.7 m of the Nussloch section bracket the LGM-period between 31.1 ± 4.1 ka and 19.8 ± 2.2 ka (Fig. 6.3). The lowermost sample, taken from the last interglacial Luvisol remains, yields an age of 122.0 ± 17.8 ka, which meets expectations and does not appear underestimated because of anomalous fading, as sometimes observed by other authors when applying IRSL-dating of

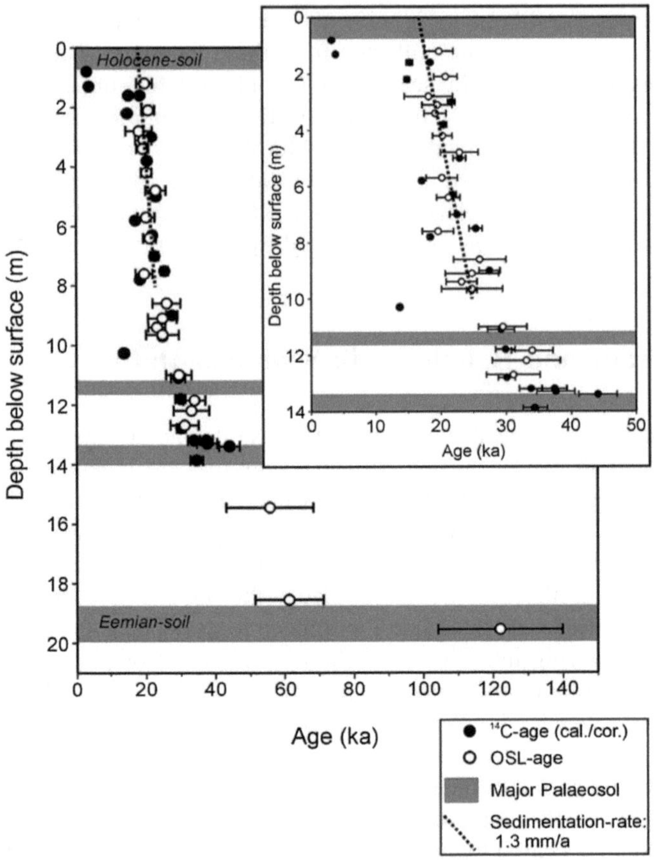

Fig. 6.3 Numeric ages versus depth below ground level at the Nussloch loess section. Apart from few exceptions, ^{14}C-ages from organic macrorest remains (wood, bone and gastropod shells) and loess organic matter (humin fraction extracted from loess fine fraction) and IRSL-ages agree generally well within error margins. The numeric ages are the chronometric backbone of the loess master section at Nussloch, which allow detailed palaeo-climatic and palaeo-environmental reconstruction (from Lang et al. 2003)

feldspars (e.g. Wallinga et al. 2000). Apart from slight age-inversion observed for the upper two samples, which is attributed to near surface disturbances during Holocene soil formation, the IRSL-ages are in agreement with the independent ages produced by AMS ^{14}C-dating (Lang et al. 2003). Such high-quality high-resolution record allows calculation of mean rates of loess accumulation (1.3 mm per year) and to interpret all other palaeo-proxies and finds appropriately.

References

Anderson A, Roberts R, Dickinson W, Clark G, Burley D, de Biran A, Hope G, Nunn P (2006) Times of sand: sedimentary history and archaeology at the Sigatoka Dunes. Geoarchaeology 21(2):131–154

Ankjaergaard C, Jain M (2010) Optically stimulated phosphorescence in orthoclase feldspar over the millisecond to second time scale. J Luminescence 130(12):2346–2355

Antoine P, Rousseau D-D, Moine O, Kunesch St, Hatté C, Lang A, Tissoux H, Zöller L (2009) Rapid and cyclic aeolian deposition during the last glacial in European loess: a high-resolution record from Nussloch. Quatern Sci Rev 28/25–26: 2955–2973

Araujo AGM, Feathers JK, Arroyo-Kalin M, Tizuka MM (2008) Lapa das Boleiras Rockshelter: stratigraphy and formation processes at a Paleoamerican site in Central Brazil. J Archaeol Sci 35:3186–3202

Barre M, Lamothe M (2010) Luminescence dating of archaeosediments: a comparison of K-feldspar and plagioclase IRSL ages. Quat Geochronol 5:324–328

Berger GW, Pérez-González A, Carbonell E, Arsuaga JI, Bermúdez de Castro J-M, Ku T-L (2008) Luminescence chronology of cave sediments at the Atapuerca paleoanthropological site. J Hum Evol 55:300–311

Berger GW (1990) Effectiveness Of natural zeroing of the Thermoluminescence in sediments. J Geophys Res 95(B8):12375–12397

Borejsza A, López IR, Frederick CD, Bateman MD (2008) Agricultural slope management and soil erosion at La Laguna, Tlaxcala. J Archaeol Sci 35:1854–1866

Bowler JM, Johnston H, Olley JM, Prescott JH, Roberts RG, Shawcross W, Spooner NA (2003) New ages for human occupation and climatic change at Lake Mungo. Nature 421:837–840

Bubenzer O, Hilgers A, Riemer H (2007) Luminescence dating and archaeology of Holocene fluvio-lacustrine sediments of Abu Tartur. Quat Geochronol 2:314–321

Bulaert JP, Murray AS, Thompsen KJ, Jain M (2009) Testing the potential of an elevated temperature IRSL signal from K-feldspar. Radiation Measurements 44:560–565

Burbidge CI, Sanderson DCW, Housely RA, Allworth Jones P (2007) Survey of Palaeolithic sites by luminescence profiling, a case study from eastern Europe. Quat Geochronol 2:296–302

Conard NJ, Malina M, Münzel SC (2009) New flutes document the earliest musical tradition in southwestern Germany. Nature 460:737–740

Cupper ML, Duncan J (2006) Last glacial megafaunal death assemblage and early human occupation at Lake Menindee, southeastern Australia. Quatern Res 66:332–341

David B, Roberts RG, Magee J, Mialanes J, Turney C, Bird M, White C, Fifield K, Tibby J (2007) Sediment mixing at Nonda Rock: investigations of stratigraphic integrity at an early archaeological site in northern Australia and implications for the human colonization of the continent. J Quat Sci 22:449–479

Derese C, Vandenberghe D, Eggermont N, Bastiaens J, Annaert R, Van den huate P (2010) A medieval settlement caught in the sand: optical dating of sand-drifting at Pulle (N Belgium). Quat Geochronol 5:336–341

Derese C, Vandenberghe DAG, van Gils M, Vanmontfort B, Meirsman E, Mees F, van den Haute P (2009) Establishing a chronology for landscape evolution around a final palaeolithic site at Arendonk-Korhaan (NE Belgium): first results from OSL dating. Mediterranean Archaeology and Archaeometry 10(4):43–51

de Moor JJW, Kasse C, van Balen R, Vandenberghe J, Wallinga J (2008) Human and climate impact on catchment development during the Holocene—Geul River, the Netherlands. Geomorphology 98:316–339

Duller GAT (1991) Equivalent dose determination using single aliquots. Nucl Tracks Radiat Meas 18:371–378

Duller GAT (1994) Luminescence dating of sediments using single aliquots: new procedures. Quat Geochronol 13:149–156

Duller GAT (2006) Comment on Human footprints in Central Mexico older than 40,000 years. by S. González, D. Huddart, M.R. Bennett and A. González-Huesca. Quat Sci Rev 25:3074–3076
Eitel B, Hecht S, Mächtle B, Schukraft G, Kadereit A, Wagner GA, Kromer B, Unkel I, Reindel M (2005) Geoarchaeological evidence from desert loess in the Nazca-Palpa region, southern Peru: Palaeoenvironmental changes and their impact on Pre-Columbian cultures. Archaeometry 47:137–158
Fanning PC, Holdaway SJ, Rhodes EJ (2008) A new geoarchaeology of Aboriginal artefact deposits in western NSW, Australia: establishing spatial and temporal geomorphic controls on the surface archaeological record. Geomorphology 101:524–532
Feathers JK (2008) Absolute dating in the Mississippi Delta. In: Rafferty J, Peacock E (eds) Time's River: archaeological syntheses in the Yazoo Basin and lower Mississippi river Valley, pp 168–181. The University of Alabama Press, Tuscaloosa
Fuchs M, Lang A, Wagner GA (2004) The history of Holocene soil erosion in the Phlious Basin, NE-Peloponnese, Greece, provided by optical dating. The Holocene 14:334–345
Fuchs M, Wagner GA (2005) The chronostratigraphy and geoarchaeological significance of an alluvial geoarchive: comparative OSL and AMS ^{14}C dating from Greece. Archaeometry 47:849–860
Fuchs M (2007) An assessment of human versus climatic impacts on Holocene soil erosion in NE Peloponnese. Quatern Res 67:349–356
Fuchs M, Kandel AW, Conard NJ, Walker SJ, Felix-Henningsen P (2008) Geoarchaeological and chronostratigraphical investigations of open-air sites in the Geelbek Dunes. Geoarchaeology 23:425–449
Fuchs M, Lang A (2009) Luminescence dating of hillslope deposits: a review. Geomorphology 109:17–26
Fuchs M, Lang A (2011) OSL dating of coarse-grain fluvial quartz using single-aliquot protocols on sediments from NE Peloponnese. Quatern Sci Rev 20:783–787
Fuchs M, Will M, Kunett E, Kreutzer S, Fischer M, Reverman R (2011) The temporal and spatial quantification of Holocene sediment dynamics in a meso-scale catchment in northern Bavaria. Holocene 21(7):1093–1104
Gibling MR, Sinkha R, Roy NG, Tandon SK, Jain M (2008) Quaternary fluvial and eolian deposits on the Belan river, India: paleoclimatic setting of Paleolithic to Neolithic archeological sites over the past 85,000 years. Quatern Sci Rev 27:391–410
González S, Huddart D, Bennett MR, González-Huesca A (2006) Human footprints in Central Mexico older than 40,000 years. Quat Sci Rev 25:201–222
Hatté C, Pessenda LC, Lang A, Paterne M (2001) Development of accurate and reliable 14C chronologies for loess deposits: application to the loess sequence of Nußloch (Rhine Valley). Radiocarbon 43:611–618
Hilgers A (2007) The chronology of Late Glacial and Holocene dune development in the northern Central European lowland reconstructed by optically stimulated luminescence (OSL) dating. Ph.D. dissertation, University of Köln, P 353
Holliday VT, Huckell BB, Mayer JH, Forman SL, McFadden LD (2006) Geoarchaeology of the Boca Nega Wash area, Albuquerque, New Mexico, USA. Geoarchaeology 21:756–802
Huntley DJ, Lamothe M (2001) Ubiquity of anomalous fading in K-feldsaprs and the measurements and correction for it in Optical dating. Candian J. Earth Sci 38:1093–1106
Hutt G, Jaek I, Tchonka J (1988) Optical dating—K-feldspars optical-response stimulation spectra. Quatern Sci Rev 7:381–385
Jacobs Z, Duller GAT, Wintle AG, Henshilwood CS (2006) Extending the chronology of deposits at Blombos Cave, South Africa, back to 140 ka using optical dating of single and multiple grains of quartz. J Hum Evol 51:255–273
Jacobs Z, Wintle AG, Duller GAT, Roberts RG, Wadley L (2008a) New ages for the post-Howiesons Poort, late and final middle stone age at Sibudu. J Archaeol Sci 35:1790–1807
Jacobs Z, Roberts RG, Galbraith RF, Deacon HJ, Grün R, Mackay A, Mitchell P, Vogelsang R, Wadley L (2008b) Ages for the middle stone age of southern Africa: implications for human behavior and dispersal. Science 322:733–735

Jacobs Z, Roberts DL (2009) Last Interglacial Age for aeolian and marine deposits and the Nahoon fossil human footprints, Southeast Coast of South Africa. Quat Geochr 4:160–169

Jacobs Z, Meyer MC, Roberts RG, Aldeias V, Dibble H, El Hajraoui MA (2011) Single-grain OSL dating at La Grotte des Contrebandiers ("Smugglers' Cave"), Morocco: improved age constraints for the middle paleolithic levels. J Archaeol Sci 38:3631–3643

Kadereit A, Sponholz B, Rösch M, Schier W, Kromer B, Wagner GA (2006a) Chronology of Holocene environmental changes at the tell site of Uivar, Romania, and its significance for late Neolithic tell evolution in the temperate Balkans. Zeitschrift für Geomorphologie N.F 142:19–45

Kadereit A, Dehner U, Hansen L, Pare Ch, Wagner GA (2006b) Geoarchaeological studies of man-environment interaction at the Glauberg, Wetterau, Germany. Zeitschrift für Geomorphologie N.F. 142:109–134

Kadereit A, Kuhn P, Wagner GA (2010) Holocene relief and soil changes in loess-covered areas of south-western Germany: the pedosedimentary archives of Bretten-Brauerbach (Kraichgau). Quatern Int 222:96–119

Kar A, Felix C, Rajaguru SN, Singhvi AK (1998) Late Holocene growth and mobility of transverse dune in the Thar desert; J Arid Envi 38:176–185

Kaiser K, Hilgers A, Schlaak N, Jankowski M, Kühn P, Bussemer S, Przegiętka K (2009, in press): Palaeopedological marker horizons in northern central Europe: characteristics of LateglacialUsselo and Finow soils. Boreas 38:591–609, DOI: 10.1111/j.1502-3885.2008.00076.x

Kim J-C, Duller GATl, Roberts HM, Wintle AG, Lee Y-I, Yi S-B (2010) Re-evaluation of the chronology of the palaeolithic site at Jeongokri, Korea, using OSL and TT-OSL signals from quartz. Quat Geochronol 5:365–370

Kind C-J (2000) Die jungpleistozänen Rinnenfüllungen von Nußloch (Rhein-Neckar-Kreis). Archäologische Ausgrabungen Baden-Württemberg 17–19

Kuper R, Kröpelin S (2006) Climate-controlled Holocene occupation of the Sahara: motor of Africa's evolution. Sci 313:803–807

Lamothe M, Auclair M, Hamzaoui C, Huot S (2003) Towards a prediction of long-term anomalous fading of feldspar IRSL. Radiat Meas 37:493–498

Lang A, Hönscheidt S (1999) Age and source of colluvial sediments at Vaihingen-Enz Germany. Catena 38(2):89–107

Lang A, Wagner GA (1996) Luminescence dating techniques in geomorphological research. Heidelberger Geographische Arbeiten 104:59–65

Lang A, Zolitschka B (2001) Optical dating of annually laminated lake sediments: a test case from Holzmaar/Germany. Quatern Sci Rev 20(5–9):737–742

Lang A, Kadereit A, Behrands RH, Wagner GA (1999) Optical dating of anthropogenic sediments at the archaeological site of Herrenbrunnenbucket, Bretten-Bauerbach. Archaeometry 41:397–411

Lang A, Hatté C, Rousseau D-D, Antoine P, Fontugne M, Zöller L, Hambach U (2003) High-resolution chronologies for loess: comparing AMS 14C and optical dating results. Quat Sci Rev 22(10):953–959

Li SH (1994) Optical dating: insufficiently bleached sediments. Radiat Meas 23:563–567

Li B, Li SH (2011) Luminescence dating of K-feldspar from sediments: a protocol without anomalous fading correction. Quat Geochronol 6:468–479

Li B, Li SH (2012) Luminescence dating of Chinese loess beyond 130 ka using the non-fading signal from K-feldspar. Quat Geochronol 10:24–31

Liritzis Y (1989) Dating of quaternary sediments by beta thermoluminescence: investigations of a new method. Analles Soc Geolog Belgique 112(1):197–206

Morwood MJ, Sutikna T, Saptomo EW, Westaway KE, DwiYaniYuniawati MW, Jatmiko Awe Due, Hadi P, Zhao J-X, Turney CSM, Fifield K, Allen H, Soejono RP, Moore R (2008) Climate, people and faunal succession on Java, Indonesia: evidence from Song Gupuh. J Archaeol Sci 35:1776–1778

Olley JM, Caitcheon GG, Roberts RG (1999) Origin of dose distributions in fluvial sediments, and the prospect of dating single grains fromm fluvial deposits using optically stimulated luminescence. Radiation Measurements 30:207–217

Olley JM, Roberts RG, Yoshida H, Bowler JM (2006) Single-grain optical dating of grave-infill associated with human burials at Lake Mungo, Australia. Quatern Sci Rev 25:2469–2474

Polymeris GS, Kiyak NG, Kitis G (2008) Component resolved IR bleaching study of the blue LM-OSL signal of various quartz samples. Geochronometria 32:79–85

Porat N, Chazan M, Grün R, Aubert M, Eisenmann V, Kolska Horwitz L (2010) New radiometric ages for the Fauresmith industry from Kathu Pan, southern Africa: implications for the earlier to middle stone age transition. J Archaeol Sci 37:269–283

Rau S, Naumann D, Barth M (2009) Eiszeit: Kunst und Kultur. Thorbecke

Rees-Jones J, Tite MS (1997) Optical dating results for British archaeological sediments; Archaeo 39:177–187

Rhodes EJ, Singarayer JS, Raynal J-P, Westaway KE, Sbihi-Alaoui FZ (2006) New age estimates for the Palaeolithic assemblages and Pleistocene succession of Casablanca, Morocco. Quat Sci Rev 25:2569–2585

Rhodes EJ, Fanning PC, Holdaway SJ, Bolton CC (2009) Archaeological surfaces in western NSW: stratigraphic contexts and preliminary OSL dating of hearths. Terra Australis 28:189–200

Rhodes EJ, Fanning PC, Holdaway SJ (2010) Developments in optically stimulated luminescence age control for geoarchaeological sediments and hearths in western New South Wales. Quat Geochronol 5:348–352

Rommens T, Verstraeten G, Peeters I, Poesen J, Govers G, Van Rompaey A, Mauz B, Packman S, Lang A (2007) Reconstruction of late-Holocene slope and dry valley sediment dynamics in a Belgian loess environment. Holocene 17:777–788

Rousseau D-D, Antoine P, Hatté C, Lang A, Zöller L, Fontugne M, Ben Othman D, Luck JM, Moine O, Labonne M, Bentaleb I, Jolly D (2002) Abrupt millennial climatic changes from Nussloch (Germany) upper Weichselian eolian records during the last glaciation. Rapid Commun Quat Sci Rev 21(14-15):1577–1582

Sanderson DCW, Murphy S (2010) Using simple portable OSL measurements and laboratory characterisation to help understand complex and heterogeneous sediment sequences for luminescence dating. Quat Geochronol 5:299–305

Schier W, Draşovean F (2004) (Eds) Vorbericht über die rumänisch-deutschen Prospektionen und Ausgrabungen in der befestigten Tellsiedlung von Uivar, jud.Timiş, Rumänien (1998–2002). PrähistorischeZeitschrift 79:145–230

Scholtz CA, Johnson TC, Cohen AS, King JW, Peck JA, Overpeck JT, Talbot MR, Brown ET, Kallindekafe L, Amoako PYO, Lyons RP, Shanahan TM, Castañeda IS, Heil CW, Forman SL, McHargue LR, Beuning KR, Gomez J, Pierson J (2007) East African megadroughts between 135 and 75 thousand years ago and bearing on early-modern human origins. Proceedings of the National Academy of Sciences 104:16416–16421

Scholtz CA, Johnson TC, Cohen AS, King JW, Peck JA, Overpeck JT, Talbot MR, Brown ET, Kallindekafe L, Amoako PYO, Lyons RP, Shanahan TM, Castañeda IS, Heil CW, Forman SL, McHargue LR, Beuning KR, Gomez J Pierson J (2007) East African megadroughts between 135 and 75 thousand years ago and bearing on early-modern human origins. In: Proceedings of the National Academy of Sciences 104:16416–16421

Singhvi AK, Lang A (1998) Improvements in Infra-red stimulated dating of poorly bleached sediments—the 'differential' partial bleach method. Ancient TL 16:63

Singhvi AK, Chauhan N, Biswas RH (2010) A survey of some new approaches in maximum age limit and accuracy of luminescence application to archaeological chronometry. MAA 10(4):9–15

Stevens M, Roberts RG, Lian OB, Yoshida H (2007) Progress in optical dating of guano-rich sediments associated with the Deep Skull, west mouth of the Great Cave of Niah. Quat Geochronol 2:330–336

Sun X-F, Mercier N, Falgueres C, Bhain J-J, Despriee J, Bayle G, Lu H-Y (2010) Recuperated optically stimulated luminescence dating of middle-size quartz grains from the Palaeolithic site of Bonneval (Eure-et-Loir). Quat Geochronol 5:342–347

Thomsen KJ, Murray AS, Bøtter-Jensen L, Kinahan J (2007) Determination of burial dose in incompletely bleached fluvial samples using single grains of quartz. Radiat Meas 42(3): 370–379. doi:10.1016/j.radmeas.2007.01.041

Tribolo C, Mercier N, Rasse M, Soriano S, Huysecom E (2010) Kobo 1 and L-Abri aux Vaches (Mali, West Africa): two case studies for the optical dating of bioturbated sediments. Quat Geochronol 5:317–323

Unkel I, Kadereit A, Mächtle B, Eitel B, Kromer B, Wagner G, Wacker L (2007) Dating methods and geomorphic evidence of palaeo-environmental changes at the eastern margin of the South Peruvian coastal desert (14°30' S) before and during the Little Ice Age. Quatern Int 175:3–28

Vafiadou A, Murray AS, Liritzis I (2007) Optically stimulated luminescence (OSL) dating investigations of rock and underlying soil from three case studies. J Archaeol Sci 34:1659–1669

Wallinga J (2008) Optically stimulated luminescence dating of fluvial deposits: a review. Boreas 31(4):303–322

Wallinga J, Murray A, Duller G (2000) Underestimation of equivalent dose in single-aliquot optical dating of feldspars caused by preheating. Radiat Meas 32:691–695

Webb S, Cupper ML, Robins R (2006) Pleistocene human footprints from the Willandra Lakes, southeastern Australia. J Hum Evol 50:405–413

Westaway KE, Morwood MJ, Roberts RG, Zhao J-X, Sutikna T, Saptomo EW, Rink WJ (2007) Establishing the time of initial human occupation of Liang Bua, western Flores, Indonesia. Quat Geochronol 2:337–343

Williams M, Pal JN, Jaiswal M, Singhvi AK (2006) River response to quaternary climatic fluctuations: evidence from the Son and Belan valleys. Quatern Sci Rev 25:2619–2631

Wilkinson K, Gerrad C, Aquilera I, Bailiff I, Pope R (2005) Prehistoric and historic landscape change in Aragón, Spain: some results from the Moncayo archaeological survey. J Mediterr Archaeol 18:31–54

Wintle AG (1973) Anomalous fading of thermo-luminescence in mineral samples. Nature 245:143–144

Wright DK, Forman SL, Kusimba CM, Pierson J, Gomez J, Tattersfield P (2007) Stratigraphic and geochronological context of human habitation along the Galana River Kenya. Geoarchaeology 22:709–728

Chapter 7
Meteoritic Impacts, Tsunamis

A meteoritic impact produces a shock wave, accompanied by a blast that creates pressure and thus heat on the surficial sediments, making them potentially datable by luminescence. Likewise, tsunami induced sediments or the sediments deposited immediately before or after a tsunami, if exposed to sunlight, can be dated (Murari et al. 2007; Huntley and Clague 1996; Sanderson and Murphy 2009).

Tsunami sediments have already been dated using luminescence techniques directly (Eipert 2004) or by dating the enclosing sediments (e.g. Reinhardt et al. 2006; Boyce et al. 2009) or both (Kennedy et al. 2007; Erginal et al. 2009; Murari et al. 2007).

The latent luminescence signal of quartz grains from small craters (ca. a few meters of diameter) may also be reset by annealing under the temperature and pressure caused by an impact (Tomasz and Stankowski 2007; Stankowski et al. 2007). For example, both impact and tsunami transport bleaching seem to apply in the case of the Chiemgau impact-induced tsunami (Sudhaus et al. 2010; Liritzis et al. 2010a), assuming that these are tsunami-like deposits and not other kind of (shock- or flood-) wave deposits.

For the impact-related OSL dates we have to take into account that pressure and temperature shock waves produced in the impact event decrease with depth. Therefore during the impact the grains in older strata should be annealed to different degrees, that is, from complete annealing to partial or no bleaching at all, provided that the lower layers have not been disturbed.

But in practice things are much more complex than the model of Stankowski et al. (2007) suggests. The decrease in pressure and temperature with depth in the contact and compression stage of impact cratering may apply on the whole. Shock propagation in inhomogeneous media like rocks is extremely complex depending on density, porosity, water content, rock boundaries, and other parameters. Consequently, the complex interference processes that shock waves normally undergo may lead to shock effects of drastic intensity fluctuations even within small volumes. Therefore, shock magnitudes depending on depth in autochthonous rocks can be quoted only statistically and not for selected samples. Even in one individual sample, shock levels may vary by orders of magnitude.

I. Liritzis et al., *Luminescence Dating in Archaeology, Anthropology, and Geoarchaeology*, SpringerBriefs in Earth System Sciences,
DOI: 10.1007/978-3-319-00170-8_7,

But matters may be even worse with regard to OSL dating of impact ejecta. Due to the peculiar trajectories of impact ejecta in the course of the excavation stage, rock material starting from far-off depths may be emplaced side by side in the ejecta deposit. Thus, the observation of Stankowski et al. (2007) and Tomasz and Stankowski (2007) of remarkably varying bleaching effects in the quartz of impact-affected samples is not surprising. On the other hand any apparent variation in bleaching should essentially appear in the cumulative dose frequency distribution of aliquots, which in the above example of Chiemgau was not observed. Despite these reservations the Chiemgau study showed that resetting was sufficient. Indeed, the locus of total doses coming from dozens of aliquots gives the impression of a rather homogeneous complete bleaching due either to pressure, or temperature of shock wave blast agents (Liritzis et al. 2010b).

References

Boyce JI, Reinhardt EG, Goodman BE (2009) Magnetic detection of ship ballast deposits and anchorage sites in King Herod's Roman harbour, Caesarea Maritima, Israel. J of Archaeological Sci 36(7):1516-1526

Eipert A (2004) Optically stimulated luminescence (OSL) dating of sand deposited by the 1960 tsunami in south-central Chile BA thesis, Carleton College, Northfield, Minnesota. http://digitalcommons.carleton.edu/gcp/3

Erginal AE, Kiyak NG, Özcan H (2009) Optically stimulated luminescence to date coastal dunes and a possible tsunami layer on the kavak delta (Saros Gulf, NW Turkey). Turk J of Earth Sci 18(3):465-474

Huntley DJ, Clague JJ (1996) Optical dating of tsunami-laid sands. Quaternary Res 46:127-140

Kennedy DM, Tannock KL, Crozier MJ, Rieser U (2007) Boulders of MIS 5 age deposited by a tsunami on the coast of Otago, New Zealand. Sediment Geol 200:222–231

Liritzis I, Polymeris G, Zacharias N (2010a) Surface luminescence dating of 'Dragon Houses' and Armena Gate at Styra (Euboea, Greece). Special issue (Keller D, guest editor). Mediterr Archaeol Archaeometry 10(3):65–81

Liritzis I, Zacharias N, Polymeris GS, Kitis G, Ernstson K, Sudhaus D, Neumair A, Mayer W, Rappenglück MA, Rappenglück B (2010b) The Chiemgau meteorite impact and tsunami event (southest Germany): first OSL dating. Mediterr Archaeol Archaeometry 10(4):17–33

Murari MK, Achyuthan H, Singhvi AK (2007) Luminescence studies on the sediments laid down by the December 2004 tsunami event: Prospects for the dating of palaeo tsunamis and for the estimation of sediment fluxes. Curr Sci 92:367–371

Reinhardt EG, Goodman BE, Boyce JI, Lopez G, van Hengstum P, Rink W, Mart Y, Raban A (2006) The tsunami of december 13, 115 A.D. and the destruction of Herod the Great's Harbor at Caesarea Maritima, Israel. Geology 34(12):1061–1064

Sanderson DCW, Murphy S (2010) Using simple portable OSL measurements and laboratory characterisation to help understand complex and heterogeneous sediment sequences for luminescence dating. Quatern Geochronology 5:299-305

Stankowski WTJ, Raukas A, Bluszcz A, Fedorowicz A (2007) Luminescence dating of the Morasko (Poland), Kaali, Ilumetsa and Tsõõrikmäe (Estonia) meteorite craters. Geochronometria 28:25–29

Sudhaus D, Ernstson K, Mayer W, Neumair A, Rappenglück B, Rappenglück M, Zeller KW, Liritzis I (2010) An ancient catastrophic tsunami event in the archaeological record around Lake Chiemsee. ANTIQUITY 84:428–439

Tomasz W, Stankowski J (2007) Luminescence dating as a diagnostic criterion for the recognition of Quaternary impact craters. Planetary and Space Sci 55:871–875

Chapter 8
Conclusions

Dating-based techniques have transformed the field of anthropology, archaeology and ancient art authentication by providing robust chronometric and analytical possibilities.

Authenticity tests on pottery items, by employing luminescence techniques, comprise by far the main percentage of the counterfeiting avocation in reputable laboratories. Ongoing research aiming to minimise measurement errors and to achieve higher accuracy while using minimal mass quantities, combined with the use of TL and OSL advanced protocols, and the exploitation of pertinent databanks, enhance the reliability of luminescence techniques and accent their usefulness as powerful means in counterfeiting detection.

In the field of OSL dating of various rocks multiple and single aliquots are used for De determination with SAR and SAAD or SAR LM-OSL. New developments include isochron IRSL of K-feldpars and TT-OSL, IR-RF, as well as, the quartz technique of limestones for surface dating.

Surface dating is becoming indispensable for dating monument architecture and lithic artifacts. Multiple sampling is needed to minimize surface destruction phenomena and careful consideration of environmental dosimetry.

The geoarchaeological examples illustrate the great potential of luminescene dating of sediments from various geomorphic process-fields (aeolian, colluvial, fluvial, tsunamis) that provide archaeological material and today, luminescence dating has become a standard tool associated to geo- and archeo-logical sciences.

Luminescence dating however is not a routine dating technique. Reliable dating results require proper studies: on the stability of the luminescence signal, on the recovery of known doses and the resetting of luminescence during at the time of burial, and on semifired conditions. In this respect each new research-region on earth and each geomorphic process-field, offers a new challenge, as the naturally inhomogeneous mineral grains and the lattice-defects pose a complex many body problem that can not be solved *ab initio*. Consequently, the samples are analyzed and then a suitable methodology that provides self consistent answers are developed and used. For geoarchaeological applications close cooperation between researchers doing the field-work and those doing the luminescence dating is crucial

I. Liritzis et al., *Luminescence Dating in Archaeology, Anthropology, and Geoarchaeology*, SpringerBriefs in Earth System Sciences,
DOI: 10.1007/978-3-319-00170-8_8, © The Author(s) 2013

already at the field-site as possibly changing hydrological conditions of a sediment over the dating-period with respective effects for the dose-rate estimation and other parameters possibly affecting the dating results have to be scrutinized carefully. Provided that such conditions of good cooperation are fulfilled, luminescence dating will most probably successfully conquer even more new research objects in geomorphology and geoarchaeology. Already luminescence dating has become an indispensible tool for revealing that climatic changes severely affected landscapes as operating spaces of past cultures and that since the Neolithic period ancient humans themselves increasingly acted on their physical environment.

Index

I. Liritzis et al., *Luminescence Dating in Archaeology, Anthropology, and Geoarchaeology*, SpringerBriefs in Earth System Sciences,
DOI: 10.1007/978-3-319-00170-8,

MIX
Papier aus verantwortungsvollen Quellen
Paper from responsible sources
FSC® C105338

If you have any concerns about our products,
you can contact us on
ProductSafety@springernature.com

In case Publisher is established outside the EU,
the EU authorized representative is:
Springer Nature Customer Service Center GmbH
Europaplatz 3, 69115 Heidelberg, Germany

Printed by Libri Plureos GmbH
in Hamburg, Germany